Birds
OF AFRICA
FROM SEABIRDS TO
SEED-EATERS

Birds
OF AFRICA
FROM SEABIRDS TO SEED-EATERS

Chris and Tilde Stuart

The MIT Press
Cambridge, Massachusetts

Acknowledgements

To the numerous past and present ornithologists,
amateur birders and naturalists who have studied and/or
taken an interest in Africa's birds and documented
their findings we say thank you. To those who will serve
to increase our knowledge of the feathered folk we say
a thank you in advance; there is still a great
deal to learn and understand.

To those who have contributed to the photographic
coverage for this book we say a special word of thanks, in
particular to John Carlyon, a true friend and a great birder.
Duncan Butchart also deserves special thanks.

To Reneé Ferreira and Louise Grantham of Southern
a special thank you. *Yol Bolsun!* To Marina Pearson, as
usual an excellent editing job, and Alix Korte, for
layout (one day we want to know how
you do it!), *merci beaucoup!*

Copyright © 1999 in text and photographs by the authors
and individual photographers
Copyright © 1999 in published edition Southern Book Publishers.
All rights reserved.

First published in the USA by The MIT Press
Published in the United Kingdom by New Holland Publishers (UK) Ltd.

Library of Congress Cataloging-in-Publication Data

Stuart, Chris
    Birds of Africa : from seabirds to seed-eaters / Chris and Tilde Stuart.
    p.   cm.
    Includes bibliographical references (p.).
    ISBN 0-262-19430-9 (hc. : alk. paper)
    1. Birds — Africa. I. Stuart, Tilde. II. Title
QL692.A1S78 1999
598'.096 — dc21   99-31341
                               CIP

Cover design by Alix Korte
Maps by CartoCom
Designed and typeset by Alix Korte
Set in Garamond light condensed 10.5/13pt
Reproduction by Hirt & Carter Repro, Cape Town
Printed and bound by Tien Wah Press
(Pte.) Ltd, Singapore

Half title page: *Lesser double-collared sunbird*
(Photo: Lanz von Horsten, ABPL)
Full title page: *Whiskered tern* (Photo: Warwick Tarboton, ABPL)
Opposite: *Greater flamingos* (Photo: Clem Haagner, ABPL)

Contents

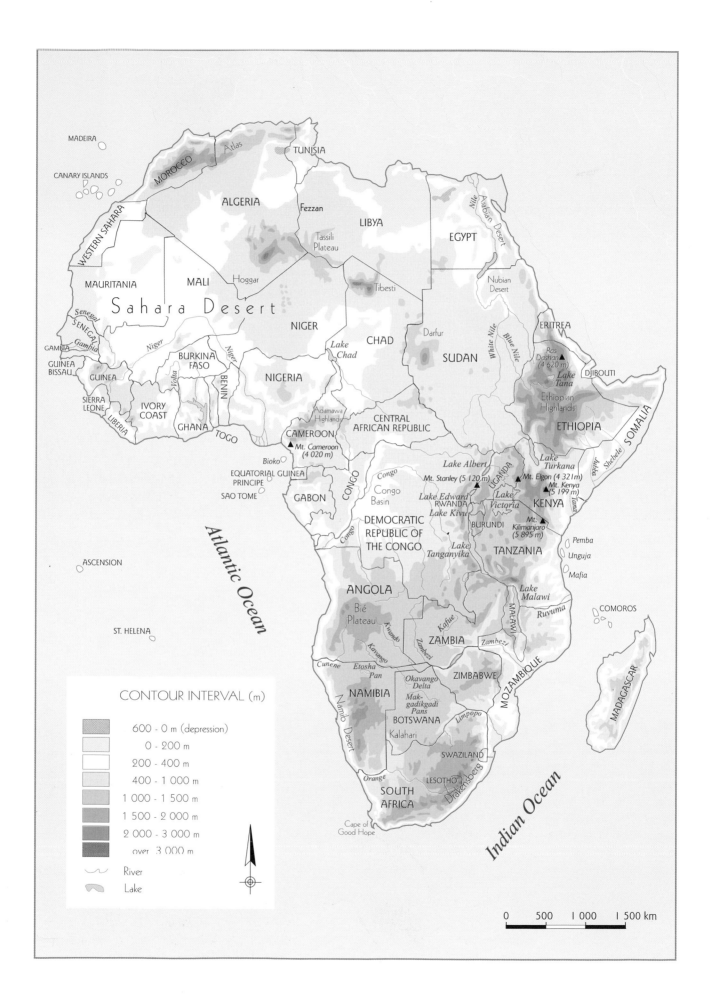

MADEIRA

CANARY ISLANDS

WESTERN SAHARA

MOROCCO

Atlas

TUNISIA

ALGERIA

Fezzan

LIBYA

EGYPT

Nile

Arabian Desert

MAURITANIA

MALI

Hoggar

Tassili Plateau

Tibesti

Nubian Desert

Sahara Desert

Senegal

SENEGAL

NIGER

CHAD

Lake Chad

Darfur

SUDAN

ERITREA

Ras Dashan (4 620 m)

Lake Tana

DJIBOUTI

GAMBIA

Gambia

GUINEA BISSAU

GUINEA

Niger

BURKINA FASO

Niger

BENIN

Volta

White Nile

Blue Nile

Ethiopian Highlands

ETHIOPIA

SOMALIA

SIERRA LEONE

LIBERIA

IVORY COAST

GHANA

TOGO

NIGERIA

Adamawa Highlands

CENTRAL AFRICAN REPUBLIC

Shebele

CAMEROON

Mt. Cameroon (4 020 m)

Juba

Bioko

Congo

Lake Albert

Mt. Stanley (5 120 m)

UGANDA

Lake Turkana

Mt. Elgon (4 321 m)

Mt. Kenya (5 199 m)

EQUATORIAL GUINEA

PRINCIPE

SAO TOME

GABON

CONGO

Congo Basin

Lake Edward

RWANDA

Lake Kivu

Lake Victoria

KENYA

Tana

Mt. Kilimanjaro (5 895 m)

BURUNDI

Congo

DEMOCRATIC REPUBLIC OF THE CONGO

Lake Tanganyika

TANZANIA

Pemba

Unguja

Mafia

ASCENSION

Atlantic Ocean

ANGOLA

Bié Plateau

Kwando

Lake Malawi

Ruvuma

COMOROS

ST. HELENA

Kafue

Kavango

Zambezi

ZAMBIA

Zambezi

MALAWI

MADAGASCAR

Cunene

Etosha Pan

ZIMBABWE

MOZAMBIQUE

Namib Desert

NAMIBIA

Okavango Delta

Makgadikgadi Pans

Kalahari

BOTSWANA

Limpopo

CONTOUR INTERVAL (m)

SWAZILAND

600 - 0 m (depression)

0 - 200 m

200 - 400 m

400 - 1 000 m

1 000 - 1 500 m

1 500 - 2 000 m

2 000 - 3 000 m

over 3 000 m

River

Lake

Orange

LESOTHO

Drakensberg

SOUTH AFRICA

Cape of Good Hope

Indian Ocean

0   500   1 000   1 500 km

1

Introduction

This book is a celebration of Africa's great avian diversity, and it is neither meant to be a field guide, nor a definitive biological text. Much is known about many of the wonderful and fascinating birds of Africa, but during the compilation of this book we have come to realise how scant our knowledge really is about much of the area's rich birdlife. What a challenge for ornithologists, present and future!

Africa lies largely within the Afrotropical Region, comprising continental Africa with the exception of the far northern extreme, as well as its associated oceanic islands, the Red Sea and the Arabian Peninsula. The portion of Africa lying north of the mighty Sahara Desert forms part of the south-western Palaearctic Region.

You could not see a cloud, because
No cloud was in the sky:
No birds were flying overhead –
There were no birds to fly.

Lewis Carroll, *Alice in Wonderland*

Where does one begin with an overview of the world's second largest bird assemblage, which spans two of the planet's great faunal kingdoms? The usual point of departure would be to give the exact total of recorded species, but this is not so easy. Taxonomists are whimsical and confusing creatures, and the difficulties are further compounded by the ever-growing taxonomic "industry" of laboratory species differentiation. Not that we are criticising the valuable work taxonomists do; just think how boring natural science would be without them ...

The African continent and its associated islands – the focus of this book – are certainly home to almost 2 400 species. Our present count comes to 2 336 but there are several species awaiting description and entry into the scientific literature, and more will follow when intrepid ornithologists manage to get into certain areas difficult of access. The latest world total is estimated at 9 946 species, but this figure is disputed. In the early years of this century ornithologists recognised almost 19 000 "distinct" species but several thousand were later reduced to subspecies rank.

Modern laboratory work, such as that undertaken by Charles Sibley, is likely to change the way we look at bird classification. The familiar order and rankings we know from our bird field guides are sure to change in time, as relationships become better understood — or perhaps more confused! It all boils down to the definition of a species, a concept that is being discussed, challenged and chewed over in scientific circles. We have not been able to track down the initial author of the following saying: *Subspecies are a matter of opinion, genera a*

BIRDS FOR AFRICA

Over the past half-century, 46 new bird species have been described from Africa. These include species discovered and described for the first time and still others separated from previously known species on the basis of clear genetic differences by using modern techniques in the laboratory. More species undoubtedly still await discovery and description.

Certain areas have dominated the "collection" of new discoveries, in particular tropical forests and other forest associations. There are, as usual, exceptions, with the Ibadan malimbe (*Malimbus ibadanensis*) having been collected in the grounds of Ibadan University in Nigeria. On the north-western shores of Lake Victoria, the Entebbe weaver (*Ploceus victoriae*) was collected and described for the first time in 1986.

Some species, once discovered and described, frustrate ornithologists by disappearing for years before new sightings are made. The Gabela bush shrike (*Laniarius amboimensis*) eluded observation between 1960 and 1992. In 1985 a single swallow flew into the Sanganeb lighthouse on the Red Sea coast of Sudan and was named *Hirundo perdita* ("lost swallow") – appropriately, as no other specimens have been found since. In 1951 the one and only known specimen of the Congo bay owl (*Phodilus prigoginei*) was collected in the Itombwe range of eastern Congo. In 1996 another owl of this species was caught and released in these mountains.

The origin of several newly discovered species is the Udzungwa range to the west of the mighty Selous Game Reserve. In 1983 the rufous-winged sunbird (*Nectarinia rufipennis*) was collected and described from this isolated fastness; then in 1991 the Udzungwa forest partridge (*Xenoperdix udzungwensis*) joined the growing list of new organisms from this recently proclaimed national park.

The new discoveries have been resident, non-migratory species, with only two exceptions: the Red Sea cliff swallow (*Hirundo perdita*) and the Mascarene shearwater (*Puffinus atrodorsalis*), described as recently as 1995. Most appear to have very restricted distributional ranges, which places many under threat.

Among the many and exciting discoveries, the Bulo Burti boubou (*Laniarius liberatus*) of Somalia is unique for the way it entered the ornithological record in 1991. First observed in the grounds of the Bulo Burti hospital in 1988, this boubou was caught in a mist-net and transported to Germany. A DNA sample was taken, as well as a few feathers, and a new species entered the African list. But for the bird this story had a happy ending, as it did not end up in a museum drawer surrounded by other unfortunates covered in dust and insecticide – one year later it was transported back to its capture site and released!

Two species of rockfowl (*Picathartes* spp.) are known from West Africa but indications are that a third species may be present in the vicinity of the Kasinga Channel linking lakes Edward and George in western Uganda. A small, unknown oxpecker (family Buphagidae) has been sighted attending buffalo in the Tai Forest of southwestern Ivory Coast in West Africa. Sadly its principal host, the buffalo, is close to extinction here, so it is deemed highly likely that this undescribed bird will follow it over the brink.

Almost unbelievably, a new species was described in 1976 from northern Algeria in North Africa: the Algerian nuthatch (*Sitta ledanti*). This constituted the first new species to be described from the ornithologically well studied western Palaearctic since 1886!

Africa is home to robins and thrushes, larks and cisticolas, a spinetail and weavers that have been sighted and reliably recorded, but are as yet undescribed.

It is certain that a substantial number of species await description by exploration, as opposed to genetic "discovery" in the laboratory. Areas exist throughout Africa where no zoological exploration has occurred in decades, even in more than 50 years. Remote high plateaux and dense forests, swamps and marshes provide an abundant habitat for birds and other biota but are often hostile to would-be ornithological explorers and naturalists. Africa's seeming predilection for long-standing civil wars are not conducive to scientific endeavour either.

matter of convenience, but species are a matter of fact. Continuing research and improving techniques may well show that a goodly number of subspecies can in fact be justifiably raised to species level – a debate that we will all be following with great interest. Actually, do total species tallies and comparisons with other continents really count? We think not and prefer to appreciate birds for themselves, not create "twitching" totals. Having said that, we do not want to denigrate the importance of reserve, regional and country lists as these are useful tools for conservation.

However, an amazing, and seemingly growing, trend today is to see birds primarily as objects of "collection". Not in the Victorian sense of shooting large numbers to then lie in serried ranks in museum drawers, soaked in arsenic, gathering dust and smelling of naphthalene, but as objects of twitching, life lists and other tallies. "What a guy, Johnnie had 2 800 twitches to his name when they buried him!" This, in itself, usually has no negative impact on the birds but they become, to a large extent, merely names to be entered on lists and tallied. The delight of observing birds – even common ones – for their own sake, for their beauty and often fascinating behaviour, seems to be a dying pleasure.

Of course, when a rarity arrives in an area accessible to large numbers of birders the disturbance can cause the avian drawcard to move on, with the ones that observed it no doubt relishing the fact that the late-comers missed their "twitch". After all, man (and woman) is a selfish beast. One could assign an increasing number of birders to a new psychiatric category: "manic twitcher"! There are a few, so we have been told by those in the know, wealthy and not so wealthy individuals who will fly thousands of kilometres in order to add one rarity to their life list. Still others bob around in boats on the very edge of a country's territorial waters in the hope of recording for the first time a petrel or shearwater to add to the nation's list. In our view the money for such excursions would be well spent if the person appreciated the bird within its environment, observed how it interacted with other species and so on, but sadly this seldom seems to be the case. Instead the prize is over

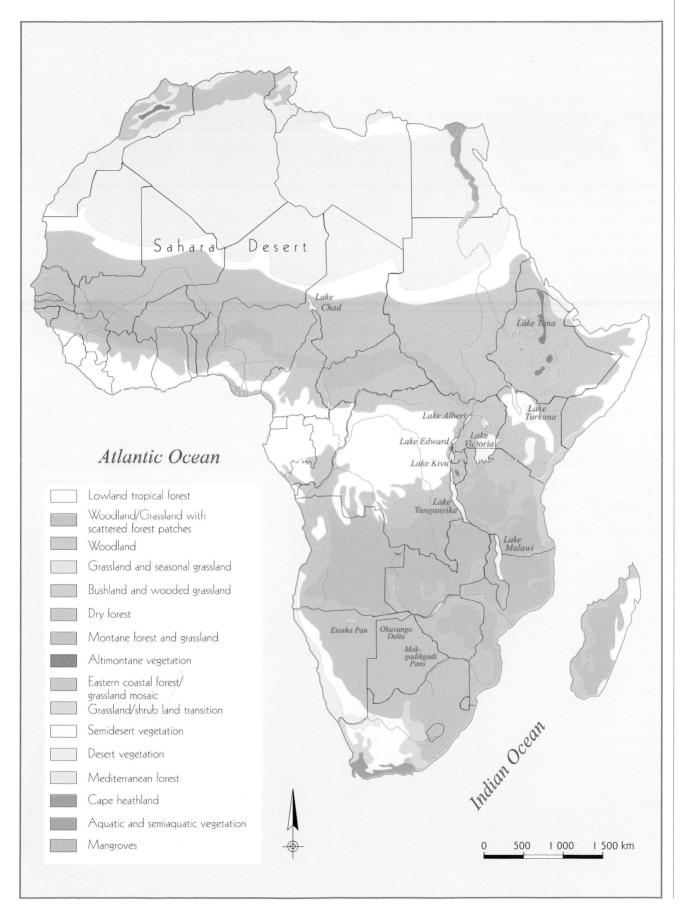

Sahara Desert

Atlantic Ocean

Indian Ocean

Lake Chad

Lake Tana

Lake Albert

Lake Turkana

Lake Edward

Lake Victoria

Lake Kivu

Lake Tanganyika

Lake Malawi

Etosha Pan

Okavango Delta

Mak-gadikgadi Pans

☐	Lowland tropical forest
▨	Woodland/Grassland with scattered forest patches
▨	Woodland
▨	Grassland and seasonal grassland
▨	Bushland and wooded grassland
▨	Dry forest
▨	Montane forest and grassland
▨	Altimontane vegetation
▨	Eastern coastal forest/ grassland mosaic
▨	Grassland/shrub land transition
☐	Semidesert vegetation
▨	Desert vegetation
▨	Mediterranean forest
▨	Cape heathland
▨	Aquatic and semiaquatic vegetation
▨	Mangroves

0   500   1 000   1 500 km

cocktails to announce that, yes, they have now added the penultimate twitch to their list.

Africa covers some 30 million km², second in size only to Eurasia, and boasts an impressive array of vast deserts, savannas, forests, lakes and mighty as well as lesser rivers, swamps and marshlands, great highlands and solitary peaks. The Indian Ocean and Red Sea coasts are graced with impressive coral reefs, and mangrove swamps occur in the deltas along these and the Atlantic coastlines. Within the Afrotropical Region are numerous oceanic islands that range in size from the diminutive to the colossus of Madagascar. All of these geographical features have resulted in the development of incredible diversities of both plant and animal life. Many bird species are therefore endemic or near endemic; others arrive as breeding or non-breeding migrants and a few are vagrants.

Africa, until some 270 million years ago, formed part of the ancient super-continent of Gondwanaland, a landmass that incorporated all the southern continents with which we are familiar today. Some 100 million years before present the new continents had all separated from each other and Africa continued its eastwards drift, until some 30 million years ago when it collided with the northern continent of Eurasia. Great natural changes took place over millions of years: forests expanded and retreated, swamplands became deserts and continental shelves were exposed and covered over again, creating environments within which Africa's rich birdlife evolved.

BIRD FAMILIES EXCLUSIVE TO AFRICA

Second only to the great Neotropical bird assemblage, the Afrotropical Region has nine bird families exclusive to it. Three of the nine "African" families spill over into south-western Arabia, which was still joined to Africa in the recent geological past and forms part of the same faunal region.

The nine Afrotropical endemic families are the Struthionidae (ostrich), Balaenicipitidae (shoebill), Scopidae (hamerkop), Coliidae (mousebirds), Viduidae (whydahs), Sagittariidae (secretary-bird), Numididae (guineafowl), Phoeniculidae (wood-hoopoes) and Musophagidae (turacos).

The ostrich has always had its principal range in the arid and savanna areas of Africa but until possibly as late as the 1960s it still clung to tenuous survival in the wildest parts of the Arabian Peninsula. But because of the actions of man, this bird – the world's largest – is now entirely confined to Africa.

In the case of the guinea-fowl, all species except one are exclusive to the African continent, with the helmeted guineafowl occurring in decreasing numbers in western Yemen and in south-western Saudi Arabia.

The hamerkop also has a limited presence in south-western Arabia.

The African continent has served as a cauldron of avian evolution. It is almost certainly here that bustards, pratincoles, sandgrouse, bee-eaters, hornbills, bulbuls, weavers, honeyguides and larks had their origins. Madagascar and the Mascarenes support a very important subfauna, with five of its bird families found nowhere else in the world. Sadly, these Indian Ocean islands also saw the demise of some families through the activities of man: the great elephant birds of Madagascar, and the dodo and solitaires of the Mascarenes.

Left: *High montane regions.*
Below left: *Sand desert.*
Below right: *High-rainfall savanna.*
Opposite page: *Grass and woodland savanna.*

Birds are the only feathered animals on earth, and they dominate the skies. Only one other group, the bats, have conquered the aerial space above the planet's crust. What is a bird? It is a vertebrate, it is warm-blooded and has a four-chambered heart, a feature shared only with the mammals, but its skeleton is closest to that of the reptiles, from which birds stemmed. Look at the scaled legs and feet of birds, the feathers which are modified scales, and what do we have — glorified reptiles, but glorious they are. Not that we wish to denigrate the reptiles; these are fine creatures on their own account.

How did the birds come to be aerial masters? Those great and long extinct flying reptiles the pterodactyls were the first flying "birds". It is probable that flight evolved as these animals ran or climbed; holding out the webbed forelimbs, they could glide in much the same way as one flies a kite into the wind. As this means of locomotion evolved, their bones became lighter and other skeletal changes took place, one of the most important being the broad and strongly keeled sternum for the attachment of the powerful flight muscles.

Some 2 000 bird species, a number of which are still alive today, have been described from their fossil remains. Birds probably had their beginnings in the Mesozoic Era some 150 million years before present. By the time of the Eocene the toothed birds had disappeared from the fossil record, and the development of many of the forms known today was well underway.

Apart from the natural changes that have taken place, the African continent has also undergone severe modification by humankind. Some of these impacts can be traced back several thousand years. Burgeoning human populations, particularly in the last 50 years, have resulted in an escalation of habitat destruction, pesticide usage and direct persecution for both food and trade. The major cause of the decline of certain bird species, and of course other biota, is the loss or dramatic modification of sensitive habitats. The rapidly dwindling forests of West Africa, and those along the East African coast and adjacent interior, are home to the majority of the continent's threatened species. If current trends continue — and it is likely that they will — we could see more and more bird species teetering towards the abyss of extinction.

Nor shall this peace sleep with her; but as when

The bird of wonder dies, the maiden phoenix,

Her ashes new-create another heir

As great in admiration as herself.

William Shakespeare, *Henry VIII*

2

Birds of the oceans

Although many of the bird species covered in this chapter are truly oceanic, a number also occur on the coastlines of these vast water bodies. A few species, such as the whiskered and white-winged terns and the grey-headed gull, are closely associated with freshwater expanses. Despite these exceptions the majority occur exclusively on the open seas and oceans around Africa and they glean their food from these waters. Because of the "difficult nature" of their chosen habitat, we know surprisingly little about many of these birds. Their movements, both migratory and between feeding grounds, their biology and their breeding habits pose numerous questions that await answers. Many oceanic birds are also notoriously difficult to identify in the field, particularly those in the group we refer to as "tube-nosed birds". Some of the most mysterious (at least to the pelagic birdwatcher!) are the prions, shearwaters and storm petrels.

The oceanic avian hordes include, in their ranks, the bird with the greatest wingspan on earth, the wandering albatross, and also the penguins, which are among the best known and loved birds. Their feeding habits are varied and interesting: the gannets plunge-hunt, the frigatebirds snatch morsels from the surface while on the wing, the puffin and penguins are underwater hunters that "fly" below the surface. A diverse range of species indeed!

Auks

FAMILY ALCIDAE

The auks are all strictly marine birds that superficially resemble the penguins of the southern hemisphere. They all have an erect stance and stocky but streamlined bodies. The four species associated with the African seaboard have dark brown or black and white plumage. All are highly adapted to obtaining their fish and other prey by underwater pursuit, using the webbed feet and the short, stubby wings to "fly" through the water. Few make landfall in Africa, staying at sea during the non-breeding season, mainly in the western Mediterranean and down the Atlantic coast of Morocco. They range in size from the barely 20 cm long little auk to the 40 cm guillemot. Overwintering flocks usually stay well away from the shore, where they undergo a moult that leaves them flightless. Having watched Atlantic puffins at one of their breeding sites on Skomer Island arriving at the nesting burrows with bills full of small sand eels, we have to admit having a soft spot for this family. ■ SEE PAGE 158.

Above: *Once in the air the gannet is an elegant flyer but take-off requires a good clear run.* **Left:** *Hartlaub's gull is restricted to the coast of south-western Africa, primarily along the Atlantic shoreline.*

Left: *The guillemot is only a vagrant to Moroccan waters.*
Below right: *A male great frigate-bird with throat-pouch inflated. This pelagic species breeds on islands and is seldom seen close to the African coastline.*
Below left: *Razorbills winter along the Atlantic coast of Morocco and into the Mediterranean as far as Algeria.*
(Photos: John Carlyon)

Frigatebirds
FAMILY FREGATIDAE

Four species of frigatebird have been recorded in oceans and seas around Africa, all in tropical waters except for the occasional vagrant that finds its way into more temperate climes.

The frigatebirds, despite their taxonomic association with such master swimmers as the cormorants, anhingas and pelicans, rarely swim or even alight on the sea. They are true oceanic aerial acrobats. On the ground they are almost as helpless as when they alight on the water, so invariably they perch, roost and nest in trees or bushes. Their wings are long, slender and pointed, and they have a long and deeply forked tail. This combination of aerodynamic perfection enables them to outfly and out-manoeuvre any other species of seabird.

It is this ability that they put to use in mobbing and chasing such species as boobies, cormorants and gulls, forcing them to drop prey held in the bill, or even to disgorge already swallowed food. Once the pirated bird has released its food the frigatebird swoops down to retrieve it before it hits the sea. Although pirating for food is quite common among frigatebirds, their main food is fish, and on occasion jellyfish, squid and even young turtles, which they snatch from the sea with their bills, the bodies remaining dry. They are also not averse to snatching beach-washed animals, as well as bird eggs and chicks from colonies – but without landing; everything is taken in flight.

The Ascension frigatebird only nests on Bosunbird Islet near Ascension Island in the tropical Atlantic, but individuals seldom approach the West African coastline. Magnificent frigatebirds only nest on the Cape Verde Islands. The greater frigatebird comes into its closest association with Africa to the northwest of Madagascar, as does the lesser frigatebird which also penetrates north into the Red Sea. ■ SEE PAGE 155.

Gannets and boobies

FAMILY SULIDAE

The gannets and boobies are large marine birds with heavy streamlined bodies, long, slender wings, prominent wedge-shaped tails and large webbed feet. All take their fish prey by diving into the sea from up to 30 m above the surface and all are armed with heavy, pointed bills. The northern gannet occurs in the waters of the North Atlantic but out of the breeding season birds move down the west coast of Africa. The Cape gannet, the only *Morus* that breeds on African inshore islands, disperses from the south-western breeding grounds into the tropics of both the Indian and Atlantic oceans but principal movements into the latter are only as far as the western bulge. The Australian gannet is a rare vagrant to South African waters.

Boobies are mainly confined to the tropics and are only rarely seen in more temperate waters, although the red-footed and brown boobies extend into the Red Sea.

One of the world's greatest sights, sounds and smells has to be a major gannet breeding colony. We have visited several of the six Cape gannet colonies off South Africa's south west coast, and I'm sure we will keep going back. The constant displaying and squabbling, and the comical landings and take-offs, supply hours of happy entertainment. After a while one becomes conditioned to the continuous cacophony, but the "fragrance" lingers in the nose long after one has left the scene. While the chicks are growing there is a constant passage of adults leaving and returning with fish and occasionally squid to feed the seemingly never-satisfied youngsters. The gannets seldom hunt far from shore, and often within sight of it. It never ceases to amaze us how the parents locate their young so quickly among this seemingly homogeneous avian mass. Of course, landing at the nest is another matter, with each arrival being mercilessly pecked until it has settled down.

The boobies were given this rather insulting name because some observers felt that they look and act stupid. This apparent stupidity manifests itself in particular on the nesting grounds where they allow close approach. As a result they have been heavily persecuted in some parts of their range. Unlike gannets, which rarely associate with ships, boobies will frequently land on their decks and rigging; they also wheel and turn across the bow catching disturbed fish, particularly flying fish.

The relatively small red-footed booby nests in trees, the other two Africa-associated species on the ground. Brown boobies breed on a number of islands off Africa, such as Principe and São Tomé in the Gulf of Guinea, and within the Red Sea and the Gulf of Aden. ■ SEE PAGE 155.

Far left and below left: *Cape gannets breed in densely packed colonies on only six offshore islands. Despite declines, their numbers are still high.*
Left: *The black head markings and blue eyes distinguish gannets from all other seabirds.*
Below: *The red-footed booby.*
(Photo: John Carlyon)

THE DILEMMA OF ISLAND BIRDS

Islands within the Afrotropical Region have a sorry record of bird extinction and looming extinction. The dodo (*Raphus cucullatus*) has become the flagship of bird extinction. This large, ungainly, ground-dwelling bird was in fact a huge, flightless pigeon which occurred only on the western Indian Ocean island of Mauritius. The rapid slide of this large bird (estimated mass: 20 kg) into extinction, probably sometime between 1681 and 1693, was the direct result of human interference. Meat-hungry European sailors killed large numbers of these huge birds, salting them down in barrels to be eaten during long sea journeys, even though the flesh was reportedly tough and not particularly tasty. Pigs and goats, introduced to the island in 1602, competed with the dodo for food, and the pigs preyed directly on its eggs and chicks. By 1644 the Dutch began to settle on Mauritius. They started

clearing the forest and planting crops. The dodo had nowhere to hide and no defences against this multi-faceted onslaught.

On the islands of Rodriguez and Réunion the solitaire (*Pezophaps* spp.) survived into the early years of the 18th century. The general belief is that there were two separate species but there is little solid evidence to back this up. The solitaire, according to our only source of information, one François Leguat, walked in an elegant fashion and had sleek plumage, and in contrast to the dodo provided excellent eating. Also unlike the dodo, the solitaire was very agile and fleet of foot, and apparently highly territorial, with males attacking intruding males and females taking on members of their own sex.

The largest birds known to have lived within the Afrotropical Region, the elephant birds (Aepyornithidae) of Madagascar, also tipped over the edge of

extinction in historical times. Some species are estimated to have topped 450 kg, making the present-day ostrich look rather meagre at 60-80 kg. The largest of all the elephant birds was *Aepyornis maximus*, the females of which laid eggs six times the size of ostrich eggs. The elephant birds are believed to have been wiped out between 500 and 700 years before now, probably as a result of two principal factors: climatic changes in Madagascar and human settlement of the island, which probably started some 1 500 years before present. The elephant birds would have provided humans with a source of meat, and were hunted to extinction.

These are the most dramatic of the island bird extinctions, but many smaller and less spectacular species have disappeared, particularly from the Mascarenes, where 25 bird species have become extinct. Still others, such as the Mauritius parakeet

(*Psittacula eques*), pink pigeon (*Nesoenas mayeri*) and Mauritius kestrel (*Falco punctatus*) came to the brink but with human intervention in the form of captive breeding programmes have been saved, for the time being.

The dodo is generally considered to be the "flagship", albeit a sad one, of extinction.
(Photo: Olaf Wirminghaus †)

Penguins

FAMILY SPHENISCIDAE

The name penguin apparently derives from two Welsh words, meaning white head. However, the name was first given to the now extinct great auk (*Pinguinus impennis*) of the north Atlantic, which had no relationship to what we call penguins. The fossil record indicates that by the Tertiary Era, some 50 million years before present, penguins were already well established within what is their present range. Although they differed in some anatomical details from present-day forms, they would be instantly recognisable as penguins. At least two early forms, from New Zealand and the South Orkneys, were larger than any form living today, although the extant emperor penguin is only 30 cm shorter than its fossil cousins.

Flightlessness in penguins is the product of more than 100 million years of evolutionary development. What they lack in flying ability they make up for in other "life skills". Although ungainly on land, in the oceans penguins are master swimmers and are on a par with seals and dolphins. No other group of birds has mastered the marine environment like penguins. Their torpedo-shaped bodies, flipper-like wings, webbed feet

and the thick insulating mat formed by the feathers ensure their oceanic supremacy. Penguins, using their flippers, literally "fly" through the water, with the webbed feet acting as rudders. On land they are at a disadvantage because with their legs positioned so far back on the body they are forced to stand upright and move either by waddling awkwardly with short steps or by hopping with both feet together (the so-called "hopping" penguins). The jackass penguin is a waddler.

The tongues and palates of penguins have spine-like extensions that enable the birds to easily grasp slippery prey such as fish and squid.

The jackass penguin is able to remain submerged for at least four minutes and descend to a depth of 130 m, but usually hunts in water depths of less than 30 m. Swimming speeds vary with age, from fledglings paddling along at little more than 5 km/h to a very respectable 19 km/h reached by fresh-plumaged adults. However, maximum speeds are usually only achieved in short bursts and commuting speeds between breeding and feeding grounds are generally less than a third of this.

Spheniscus penguins, which include the jackass, feed primarily on pelagic, shoaling fish such as pilchards and anchovies. Unlike penguins of other genera, these birds do not have plain white underparts but have bold black and white stripes along the side of the head and breast. This patterning makes them more visible to underwater predators, such as Cape fur seals (*Arctocephalus pusillus*) and sharks, but it increases their success at fish hunting by serving to manipulate fish behaviour. Experiments have shown that when captive fish are exposed to models with the boldly striped pattern of the jackass penguin, the shoals break up more often than when shown non-patterned models. Jackass penguins hunt in small flocks, circling fish shoals, with individual birds alert to dash out and take fish attempting to break away from the shoal. By contrast, young birds lack the bold markings, because at this stage of their lives they swim more slowly than adults and camouflage is essential to limit predation. Their lack of speed also means that they are unable to catch fast-swimming pelagic fish, and instead they rely on solitary prey fish which they need to approach with greater stealth.

Penguins are found only in the southern hemisphere, with the major concentration of species occurring in the cold waters along the Antarctic coastline and on mid-oceanic islands. Africa has only one resident and endemic species, the jackass (or black-footed) penguin. Three species are recorded in the far south as rare vagrants from the colder southern oceans. The large king penguin has been recorded only once from Africa, and it has been suggested that this might have been a bird transported by ship and then released. The macaroni penguin has only been recorded four times on the southern shore of South Africa, but the rockhopper penguin is a regular, albeit rare, vagrant to the southern coastline.

The medium-sized jackass penguin, with a total length of some 60 cm and up to 3,5 kg in mass, breeds on small inshore islands and in a few cases on the mainland. This colonial species has suffered perhaps as much as a 75% population decline in this century. Initially, the threat stemmed from the uncontrolled harvesting of eggs, a staggering 14 million in the first 30 years of the 20th century. Harvesting of penguin eggs continued until 1969, when legislation was enacted to protect these unique marine birds.

Present estimates of jackass penguin breeding pairs range from 500 000 to one million, nesting on 20 offshore islands and two mainland sites. Several factors, other than past egg harvesting, have influenced penguin numbers, but to some extent certain measures have been implemented to help stem the decline in their population. Guano for use as fertiliser was, and still is, collected at seabird nesting and roosting sites off the coast. Apart from causing direct disturbance of the penguins and other birds, the removal of the thick layers of guano has in some cases deprived the penguins of burrowing locations in which to lay their eggs and raise their chicks. Protection has now been afforded to those sites used by penguins for nesting. It has been estimated that the present jackass penguin population consumes some 45 000 tons of pelagic fish each year, in contrast to the hundreds of thousands of tons of this oceanic bounty harvested by man.

During the 1950s and 1960s commercial fishing fleets off the south-west African coast overfished stocks of pilchards (*Sardinops ocellata*), then the most important component of the penguins' diet. Penguin numbers went into serious decline, but over time the birds shifted their diet away from the scarce pilchards and adapted to hunting abundant non-commercial species. ■ SEE PAGE 155.

Above: *Jackass penguins returning to the colony.*
Right: *Jackass, or black-footed, penguins preening after returning from the open sea. This is Africa's only resident penguin species.*

Kittiwakes visit the waters off north-western Africa during winter. (Photo: John Carlyon)

Skuas and jaegers, gulls and terns

FAMILY LARIDAE

Some authorities place each of the skuas, gulls and terns in their own family, but others lump them all into the family Laridae. All African species are primarily aquatic, mainly over the oceans, seas and their coastlines, but also over the extensive inland waters of the continent. A few species are pelagic outside the breeding season, not venturing near land during that time. Virtually without exception they are strong, skilful fliers with long wings. The legs are short to very short (in some terns) and the feet are well webbed as an aid to swimming, which many species do. The bills are fairly short but usually heavy and powerful, although those of the terns tend to be more slender and lacking a terminal hook. Plumage coloration is predominantly black and white or grey and brown, depending on the species. This family includes some of the birds best known to non-birders: the ubiquitous seagulls. The family has a wide range of foraging techniques; some members scavenge, others actively fish and hunt, and opportunists combine both. There are ground-hunters and plunge-hunters, such as many of the terns which plunge into the water in much the same manner as gannets, snatching small fish and other animals on or close to the surface. There are also those that rely on other species to provide them with a large percentage of their food requirements. Although a few are African endemics, the vast majority are not restricted to the African continent, arriving as non-breeding migrants; 21 gulls, one kittiwake, six skuas and jaegers, and no less than 24 terns are associated with Africa.

SKUAS & JAEGERS

No skuas or jaegers breed in African waters, but they move north or south through the region to escape the ferocious polar winters. All are gull-like in overall appearance with browns, greys and whites dominating.

The name "skua" has Norwegian origins, and "jaeger" German and Dutch, in reference to their hunting skills. The largest member of this group, appropriately named the great skua, looks superficially like a massive immature gull with its brown-mottled plumage. However, within the species there is considerable variation in size. Like all skuas it is highly pelagic and seldom seen close inshore. It is the only bird that breeds in both the Antarctic and Arctic belts.

While at sea off the African coast, and elsewhere, great skuas obtain most of their food by harassing other birds, thus forcing them to drop and regurgitate their prey. They may grasp the wing or tail of the victim and drag it seawards until it relinquishes or regurgitates its hard-earned meal. This pirate method of obtaining food is called kleptoparasitism. Great skuas also actively fish and forage on the sea's surface, in much the same way as gulls. They also attack and kill other birds, such as petrels, prions and smaller gulls, which they eat. Although usually observed in small numbers in the wintering grounds, hundreds may gather around fishing fleets or some abundant food source.

The main wintering range of the parasitic jaeger lies off the south-western coast of Africa and is closely allied to the food-rich Benguela Current. Another species that concentrates in these rich, cold waters off Africa is the long-tailed jaeger, with its long tail-streamers reminiscent of tropicbirds. The south polar skua is a rare vagrant to the southern and east African ocean currents. ■ SEE PAGE 158.

GULLS

Gulls are usually the most visible and abundant of the medium to large, heavy-bodied birds along African shorelines, although some are predominantly pelagic in their overwintering grounds. Many gull species associate with people and their activities, whether on holiday beaches or at fishing villages, harbours and major ports. Several species use artificial structures as alternative nesting sites to cliffs and beaches, not always to the joy of the naked ape! A number of species have no scruples as to where they seek their daily food: large numbers congregate around sewerage outlets, rubbish dumps and land fill sites, squabbling over the waste spawned by humans. Gulls are true opportunists when it comes to finding food; apart from scavenging, they are active predators on the sea, inland waters, beaches and shorelines, even following ploughs on agricultural land for insects and worms turned up by the blades. There are those that exhibit what we call indirect tool-use: carrying molluscs, particularly bivalves, aloft and dropping them until they break, allowing access to the edible interior. Unfortunately, this skill is not perfected and many birds do not differentiate between rock and soft beach as a suitable "tool" for breaking open the meal ... There are also a few species that are adept at aerial hunting, and we have watched wintering black-headed gulls hawking for flying termite alates.

It takes several seasons and moults before juvenile birds take on the handsome, usually two-toned, plumage of adults — even as long as four years in the larger species. These intermediate stages cause much identification confusion. A factor that makes many gulls stand out from other birds is their generally raucous and frequent calls, most of which cannot be described in the remotest sense as being musical; nevertheless, they have a certain charm.

Most gulls nest in fairly dense colonies but pairs are monogamous and defend the individual nest site. They do not hesitate to "dive-bomb" intruders, human or otherwise.

The vast majority of gulls fall within the genus *Larus*. They range from small to large in size. Many gulls hold the attentions of taxonomists, some of whom favour a lumping and others a splitting of species. The disputes are caused by the lesser black-backed and yellow-legged or herring gull complexes.

Just eight gull species breed in Africa, and most have limited nesting grounds. Breeding by sooty and white-eyed gulls is restricted to islands in the Gulf of Aden and in the Red Sea. Hartlaub's gull, a southern African coastal endemic, breeds in colonies up to 1 000 pairs strong, mainly on offshore islands but also at a few sites on the mainland. Another south-western African breeding endemic is the large, dark-backed kelp gull, with perhaps 12 000 breeding pairs nesting at more than 50 locations. Most nesting sites are located on inshore islands but a few mainland sites exist, although these are vulnerable to disturbance and predation. One of Africa's most abundant breeding gulls is the grey-headed gull. Most breeding sites are located on inland waters and several number more than 1 000 pairs — a noisy situation indeed. Where the grey-headed and Hartlaub's gulls occur together they may be confused by observers. The slender-billed gull breeds in West and North Africa, among other places, in relatively small colonies on

Left: *Herring gulls breed along the Mediterranean coastline of Africa.*
Below: *The Arctic skua is a fairly common oceanic species that frequently comes close inshore.*
Below right: *Kelp gulls are rarely encountered inland from the coast.*

islands. Audouin's gull is a near endemic to the Mediterranean, where it usually feeds offshore and often at night, flying just above the surface and lunging at fish. It also scavenges, consumes large quantities of insects and takes many small birds during the annual migration cycle. The herring gull, restricted when breeding to the northern hemisphere including the North African coastline, also occurs elsewhere as a migrant during the non-breeding season. Several apparent subspecies (some regard them as distinct species) are seen on the African shoreline; differences are minimal and open to debate.

The migrant Sabine's gull breeds in the far north from Alaska to Siberia but a large percentage of the population moves down off the west coast of Africa to overwinter on average 40 km from the shore of Namibia and South Africa. Although usually concentrated in small flocks, up to 2 000 birds may associate with pelagic fishing fleets. Within their northern breeding grounds the black-headed and lesser black-backed gulls tend to be more marine and coastal than in their overwintering grounds. Substantial numbers of lesser black-backed gulls penetrate along the entire length of the Nile River, the Great Rift Valley and the Niger River, as well as much of the continent's coastline. ■ SEE PAGE 158.

TERNS

Terns are arguably the most elegant and streamlined of aquatic birds. They are small to medium-sized, in most cases smaller and more lightly built than gulls, with slender, pointed wings and usually a forked tail. Particularly in non-breeding plumage a number of species are notoriously difficult to separate. With a few exceptions plumage coloration is a combination of greys and white. Although their feet are webbed, terns seldom swim and despite short legs they manage to walk well, but usually over very short distances. Terns hunt mainly by hovering, then plunging to take prey from the water surface, even diving below the surface. Predominantly freshwater-inhabiting terns also hawk for insects above, and around, water bodies.

Although many terns are non-breeding migrants to African shores, several breeders are of international importance. Only one species is an African endemic: the Damara tern. One of the world's largest breeding populations of the roseate tern is located in Kenya but numbers vary from year to year. Other terns with important breeding populations include the royal, with thousands of pairs breeding at a few sites on the western bulge. Great crested and lesser crested terns breed in the south and east. Breeding colonies of the sooty tern can reach massive proportions, with up to 200 000 pairs having been recorded on islands in the Gulf of Guinea. Important populations of the black noddy are also found there. Terns at breeding colonies can be extremely aggressive in defence of eggs and chicks, often not hesitating to attack human and other intruders. With their sharp-pointed bills they can easily draw blood. Despite these efforts tern colonies are subjected to many threats, including the uncontrolled harvesting of eggs, predation by introduced rats and high levels of disturbance that may cause the adults to desert. It is known that some, if not all, terns are long-lived, with 20 and more years not being unusual.

Like gulls, terns have two annual moults, and it takes from two to three years for the young to take on full adult plumage.

Above left: *Sooty gulls in flight.*
Above: *Arctic tern.* (Photo: John Carlyon)
Left: *White-winged tern in non-breeding plumage.*
(Photo: Nigel Dennis, ABPL)
Far left: *Slender-billed gulls are known to breed in small numbers in north-western Africa, and non-breeding migrants occur along the coast of the Horn of Africa and inland to Lake Turkana.*

Whiskered tern parent and young at nest, Wakkerstroom, South Africa. (Photo: Warwick Tarboton, ABPI.)

Even in adult form, particularly in non-breeding plumage, many terns are notoriously difficult to identify in the field. Nevertheless they are wonderful, elegant birds to observe – after all does it really matter which species you are observing? The common tern, with an almost international range, has been recorded around the entire length of the African coastline and even at a few inland locations. It is one of the most frequent subjects of avian ringing exercises – a considerable weight of aluminium is carried by these birds!

One of the avian world's migration gold medals must go to the Arctic tern, a wanderer of unequalled scope. Breeding in the northern hemisphere, across North America, Greenland, northern Europe and eastwards to the Bering Sea, they migrate to spend the winter in southern waters, many along the Antarctic continent. Two principal flight paths are followed, one down the west coast of the Americas and the other along Africa's western seaboard.

Apart from the many coastal and marine terns, there are several that also frequent inland waters, and still others that seldom venture to the coast from the interior. The gull-billed tern, sometimes allocated its own genus *Gelochelidon*, occupies coastal margins, particularly where there are mangroves, but also penetrates deep inland along the Nile as far south as Lake Tanganyika. It occurs widely in West Africa, including along the entire length of the Niger River. These terns feed mainly on insects, scooping them off the water's surface as well as hawking them in the air. We have observed them on the

Kenyan coast scooping up fiddler crabs, and in southern Arabia taking small jellyfish from near the surface in shallow water. The very large and distinctive Caspian tern occurs around much of the African seaboard and closely follows the major rivers where it dives for fish up to almost a quarter of a kilogram. Little terns are mainly coastal but frequently penetrate along the Niger and the Benue rivers.

The whiskered and white-winged terns both have extensive African ranges but only the whiskered has a resident as well as migratory population. The white-winged tern is a non-breeding migrant. Although insects make up the bulk of their diet, both species will also readily take small fish, frogs and tadpoles. They have been recorded as plucking caterpillars from the ground and vegetation, and the white-winged is known to follow tiger fish, catching the tiny fish that leap out of the water in their attempts to escape the predator.

Some terns lay but one egg, usually in a shallow scrape or natural hollow with little lining. The whiskered tern constructs a bulky floating structure of plant material, of which much lies below the surface.

There is some doubt as to whether the white, or fairy, tern does occasionally come close to the African mainland from its oceanic island breeding grounds, but its unique "nesting" behaviour deserves a mention. The egg is laid, precariously, directly on a branch with no nest, and on hatching the single chick remains for several days on its lofty – for a tiny chick – perch before it either jumps or falls off. ■ SEE PAGE 158.

White-tailed tropicbird with a large chick. (Photo: John Carlyon)

Tropicbirds

FAMILY PHAETHONTIDAE

Tropicbirds are highly aerial seabirds with predominantly white plumage, and with variable amounts of black depending on the species and age. In adult birds the dominant feature is the long, slender tail-streamers, which are red in the red-tailed tropicbird but white in the other two species. It was these tail-streamers, once likened by sailors to the marlin spikes they used, that gave them the name of bosunbird.

These are highly pelagic birds which out of the breeding season may wander hundreds of kilometres away from their nesting islands. They feed mainly on squid and fish, using the fishing technique they share with gannets: half folding the wings, plunging into the sea and then briefly remaining on the surface before taking wing again. A special cushion of air-filled cells under the skin of the neck and chest serves to protect the birds when they dive at speed into the water. Although large numbers of tropicbirds gather on their breeding islands, they usually forage singly.

All three tropicbird species have been recorded in African waters, principally in the tropics but also as stragglers off the cooler coastlines. Both the red-billed and white-tailed tropicbirds occur and breed in the Indian and Atlantic oceans but near Africa the red-tailed tropicbird is found only in the Indian Ocean. The red-billed tropicbird occurs as far north as the Red Sea where it breeds. ■ SEE PAGE 155.

TUBE-NOSED BIRDS

ORDER PROCELLARIIFORMES

Of the approximately 116 "tube-noses" known, at least 53 species have been recorded in African waters. All species are closely related and all share the same bill structure. In all cases the nostrils form short tubes that extend onto the upper surface of the bill; in most cases they are linked on the culmen but in a few species, notably the albatrosses, they open separately on each side of the bill, or the culminicorn section. The bill is strong, terminally hooked and comprises several horny plates. Above each eye lies a well-developed gland that excretes excess salt from the seawater which of necessity they must drink.

Tube-nosed birds range in size from the bird with the greatest wingspan on earth (in some specimens more than 3,5 m), the wandering albatross, to some of the storm petrels which are little larger than some robins.

All members of this order are highly pelagic: they spend much of their lives on and above the open seas, only coming to land to breed. Nearly all species breed on oceanic islands, and the vast majority lay their single white egg in an underground burrow. The exceptions are the albatrosses, which construct nest mounds on the ground, and the fulmars, which lay their eggs on cliff ledges.

Another peculiarity of many members of this order is that they are active at their breeding grounds only at night, and their incubation and fledgling periods are remarkably long.

Newly hatched chicks are fed a yellowish stomach oil by the parents, and later regurgitated fish, squid or krill.

The tube-noses have no near relatives in the other bird orders but we do know that their ancestors were around at least 60 million years before present, and possibly even earlier.

Albatrosses

FAMILY DIOMEDEIDAE

Of the 11 species of albatross and mollymawk occurring over the oceans surrounding Africa, all are large with long, slender wings designed for gliding in the strong winds that buffet the southern oceans.

The name mollymawk for smaller albatrosses derives from the Dutch word *mallemowk*, meaning foolish gull – not very elegant for such fine aerial masters! When the Portuguese sailing ships first started venturing into the south Atlantic Ocean in the 15th century they observed large black and white birds with long and slender wings, which they called alcatraz. English sailors are said to have corrupted this to albatross – apart from the first two letters we don't see the connection but this is what the linguists would have us believe. In the past, sailors, ever short of fresh meat, would kill albatrosses follow-

Above: *The fulmar is a vagrant to Moroccan waters.*
(Photo: John Carlyon)
Left: *Herald petrel nesting on Round Island, near Mauritius.*
(Photo: Anthony Bannister)
Far left: *Wedge-tailed shearwater chick at its nest under a rock overhang, Round Island.*
(Photo: Anthony Bannister)

ing their ships. Also, when landfall was made on island breeding grounds, eggs and fat chicks were harvested. In the late 19th century albatrosses were slaughtered in large numbers for their feathers. Wings were used to construct the rather grotesque ladies' hats that were in fashion at that time and body down was sold as "swan's down" for filling mattresses and pillows. To what extent this affected the albatross and mollymawk numbers around the African coastline is not clear but as hunting was concentrated around the breeding islands it surely had a major impact on overall numbers.

The albatrosses, highly accomplished and evolved, are dynamic soaring birds. Air currents are essential for these magnificent birds to maintain sustained flight. When wind speeds are low, albatrosses normally settle on the water surface until stronger winds begin to blow. The calm conditions that prevail in the tropics of the Atlantic and Pacific oceans generally act as a barrier preventing the southern albatrosses and mollymawks from regularly penetrating into the northern hemisphere. The centre of albatross distribution lies in the southern oceans in the "roaring forties" and "furious fifties" – a reference to those latitudes' renown for their regular gale-force winds and stormy seas – an albatross paradise! Because of their large body size albatrosses require fairly long runs into the wind, both on land and water. Island nesting sites are usually on slopes in order to facilitate "launching".

Their superb gliding ability allows parent birds to travel great distances between the breeding sites and the principal feeding grounds. In some cases parent birds have been found more than 3 000 km away from their chicks, and on their return to regurgitate almost 2 kg of squid and stomach oil. Squid forms the bulk of the diet of most albatrosses, but they also regularly scavenge around fishing vessels. In Antarctic waters several albatross species also take krill, but their poor diving ability limits them to snatching these small crustaceans from close to the surface.

No albatross or mollymawk species breeds on any island close to the African mainland but several species are commonly sighted in waters along and beyond its continental shelf. Clear views of any albatross from the African shoreline are very rare; those birders intent on seeing these aerial masters need to take to boats. As with nearly all things, there is an exception: the black-browed mollymawk. This albatross not infrequently comes close inshore and may gather in hundreds on commercial fishing grounds, gleaning offal from the boats. Although the yellow-nosed mollymawk will also gather in numbers around vessels on the pelagic fishing grounds, the other albatrosses and mollymawks generally occur in smaller numbers or are rare wanderers off the African coastline. The latest addition to the African list is a single, verified sighting of Buller's albatross (*Diomedea bulleri*). ■ SEE PAGE 155.

A large group of yellow-nosed mollymawks, Cape gannets and petrels near a fishing trawler. (Photo: Alan Wilson)

Below: *The broad-billed prion is somtimes subject to "wrecks" that may invlove thousands of birds.* (Photo: Peter Steyn)
Left: *A wandering albatross feeding its chick, Marion Island.* (Photo: Prof. Rudi van Aarde)
Below left: *A wedge-tailed shearwater on her egg.* (Photo: Anthony Bannister)

Petrels, prions, fulmars and shearwaters

FAMILY PROCELLARIIDAE

Of the 17 petrels, five prions, two fulmars and 10 shearwaters recorded in the seas around the African continent, few are ever seen close inshore. Members of this rather diverse pelagic family range in size from the Antarctic, or southern, giant petrel, with a wingspan that can exceed 2 m, to the thin-billed prion with a 56 cm wingspan. With a few exceptions, the species in this family are difficult to identify. Petrels of the genus *Pterodroma* are particularly difficult to separate.

Petrels and shearwaters have the nostrils encased in tubes joined on top of the base of the bill. The so-called fulmarine petrels (e.g. *Macronectes*) have very long nasal tubes. The so-called gadfly petrels (*Pterodroma*, *Halobaena* and *Bulweria*) are characterised by short heads, short black bills and long, slender wings. Petrels breeding on Africa-associated islands include the soft-plumaged and Bulwer's, but most that are found in African oceanic overwintering grounds breed on sub-Antarctic islands.

There is some controversy over the number of species of prion, with different authorities favouring three, five and six species. Prions all have bluish-grey plumage and are frequently referred to as whalebirds. They are specialised feeders with rather broad bills. Two rows of "filters" or lamellae on the palate filter plankton from seawater, which is obtained from the surface or by diving.

Shearwaters are all migratory to a greater or lesser extent, with a few species making epic journeys. They take their name from their habit of skimming just above the surface of the water, dipping and soaring over often mountainous waves with almost motionless wings. The coloration of shearwaters is either all dark, or dark and white, and identification is limited by the experience and ability of the human observer. Although knowledge of the species' range can be useful when pinpointing an identity, many species are great migrants and wanderers; they can, and do, turn up off their normal "beaten track". Some species occur in great numbers both at the breeding sites and the overwintering grounds. Cory's shearwater breeds on many islands in the Mediterranean, as well as on the Canary and Cape Verde islands in the Atlantic Ocean off the African coastline. Those that breed in the Mediterranean move in from the Atlantic at a rate of some 3 600 birds per hour through the Strait of Gibraltar. At the end of the breeding season when the adults and young birds depart for their overwintering grounds in the open ocean, they stream through the Strait in October-

November at an estimated rate of 26 272 each day! Although Cory's shearwaters disperse widely, many overwinter off the coasts of Namibia and South Africa, with some dispersal into the Indian Ocean. Other shearwaters breeding on Africa-associated islands include the Manx in the Mediterranean and on Madeira, and the little shearwater on the Canary and Cape Verde islands. ■ SEE PAGE 155.

Sheathbills

FAMILY CHIONIDIDAE

The sheathbills, only the greater having been recorded as a solitary vagrant from cold Antarctic waters to South Africa, are strange-looking birds believed to form the link between the waders and gulls. Their name is derived from the horny,

Above: *Fulmar flying over nesting ledge.* (Photo: Brendan Ryan)
Left: *Antarctic petrel.*
(Photo: Peter Steyn)
Far left: *The southern giant petrel, the size of a mollymawk, breeds on oceanic islands in the cold southern oceans.*
(Photo: Peter Lillie)

Above: *Wilson's storm petrel is a "master" migrant, moving between the breeding islands and through the oceans.* (Photo: Peter Steyn)
Right: *The lesser sheathbill is only known on a handful of subantarctic islands. Although not strictly falling within South African waters, some of these islands are under South African jurisdiction. Only the greater sheathbill has been recorded from the mainland.* (Photo: Peter Steyn)

saddle-like sheath that covers the base of the upper bill. The greater sheathbill has all-white plumage. It forages around bird breeding colonies and seal rookeries, including their droppings in its diet. ■ SEE PAGE 158.

Storm petrels

FAMILY OCEANITIDAE

Of the 20 species of storm petrels in eight genera, 10 have been recorded in African waters. These are the smallest of the tube-nosed birds, some with a wingspan of less than 25 cm. They are broadly divided into two groups: species of the northern hemisphere have short legs, more or less forked tails and longer, more pointed wings; southern hemisphere species have shorter, rounded wings and long legs with which they skitter over the water with raised wings and feet touching the surface. The name petrel is therefore said to derive from the biblical account of St Peter walking on water. The "storm" part of their common name derives from an old mariners' superstition that these small birds are present when a storm is brewing. Another name that has been bestowed upon them is "Mother Carey's chickens" – although not exactly clear how, it is believed to come from *Mater Cara*, the biblical Divine Virgin who is the guardian of sailors.

Although some storm petrels remain in limited oceanic areas, most are migratory. An example is Wilson's storm petrel, which migrates from its Antarctic breeding grounds northwards to the sub-Arctic oceans and back again, an impressive feat for a bird of less than 50 g. By May they reach the Gulf of Aden and the Red Sea, where flocks numbering in the thousands may gather. By July they are heading back, past the Indian subcontinent, on their way to their breeding grounds.

Petrels feed on a mix of crustaceans, including krill, as well as squid, fish and offal. Some species are more specialised than others. Some gather in great numbers, both at the breeding and overwintering grounds. They nest in burrows and in common with the rest of this order they lay a single egg. The nesting colonies are only active at night; during the day one parent incubates the egg. After hatching the chick is fed by both parents returning from the sea after nightfall. Only when the youngster is fully fledged, which can take up to 50 days in some species, do the parents desert it. Hunger eventually forces it to leave the burrow, at night, and make its own way on the stormy oceans and seas. Some species, such as Wilson's storm petrel, frequently follow ships to glean scraps, whereas others rarely associate with these human food sources.

The European storm petrel breeds on a number of Mediterranean islands, including off Tunisia, but the status of these African colonies is not known. The white-faced storm petrel breeds on the Canary and Cape Verde islands off the African coastline, and colonies of the band-rumped storm petrel are known on the Cape Verde islands, St Helena and Madeira. ■ SEE PAGE 155.

3

Birds of inland waters and coastline

Wetland habitats provide permanent and seasonal homes for the vast majority of species covered in this chapter, but as with nearly all things there are exceptions. For example, some of the plovers, lapwings and storks favour drier habitats, and it is not unusual to see such species as grey and particularly black-headed herons hunting some distance from water. However, nearly all members of this diverse mix rely on watery habitats for refuge, food and "living quarters". Included in this large assemblage are highly specialised feeders, such as the flamingos; many others are generalists and opportunists. Some birds obtain their food at the very edge of the water, some probe with long bills in mud and sand, others swim and yet others dabble. A number are adept and efficient divers and underwater hunters, for example cormorants, anhingas and grebes.

Avocets and stilts

FAMILY RECURVIROSTRIDAE

Of the seven recognised species worldwide just two, the avocet and black-winged stilt, occur in Africa. Neither is exclusive to the continent and both have wide distributional ranges.

Both have black and white plumage, and each has an obvious anatomical peculiarity. The stilt as its name implies has enormously long legs, and the avocet a long, upwardly curved bill. Stilts use their long, thin and straight bill to probe for invertebrates in mud. The scimitar-shaped, flattened and lamellae-equipped bill of the avocet is swept, slightly parted, from side to side in the shallows either on the surface or on the bottom. Avocets also readily swim and feed on the surface in deeper water. When an avocet feeds on a muddy bottom its swinging bill leaves distinctive marks, as do the partially webbed feet. Stilts also have some webbing between the toes.

Although stilts and avocets are commonly seen in pairs or small flocks, on occasion they gather in hundreds. Some years ago we were camped at the edge of Sowa Pan in Botswana, which at the time was covered by just a few centimetres of water. We watched more than 500 avocets flying and foraging against the backdrop of a dust-filtered setting sun, as only Africa can produce.

Both species will suddenly arrive at seasonally flooded pans that are surrounded by otherwise totally inhospitable habitat, to feed on the abundant rain-liberated organisms. The question comes to mind each time we see these wetland birds on temporary arid-area pans, how did they know?

Although pairs may nest in isolation, both species usually nest in loose colonies. ■ SEE PAGE 158.

Above: *The eastern white pelican has a large wing surface, required to lift its heavy bulk (up to 15 kg) into the air and keep it there.*
Left: *The black-winged stilt; as it name implies, this bird has enormously long legs.* (Photo: Richard du Toit)

Crab plover

FAMILY DROMADIDAE

This is a very unusual bird that we have come to know well from our work in Arabia, Zanzibar and on the Kenyan coast. It is a fairly large, exclusively coastal species, black and white and with long legs and a strong, pointed tern-like bill. Crab plovers usually associate in small flocks but up to a few hundred may gather. They are colonial nesters. Their nesting habits are unusual: they dig nesting burrows in level sand or low sand banks. The female lays a single egg which is very large for the size of the bird. The eggs are white as is typical of hole-nesting birds, and the young remain in the burrow where they are fed by both parents. The principal breeding grounds are located around the Arabian Peninsula. Only Somalia is known to accommodate nesting colonies in Africa.

We have watched crab plovers feeding during the day and on moonlit nights, on mudflats in mangrove swamps, on low sandbanks at low tide, as well as on coral reefs. ■ SEE PAGE 158.

Right: *The dark-breasted form of the great cormorant is not found in southern Africa.*
Far right: *Crab plovers, placed in their own family, are unusual in a number of ways, including the fact that they nest colonially in burrows on sandy beaches.*

Cormorants

FAMILY PHALACROCORACIDAE

Whenever we think of cormorants two visions immediately spring to mind: the great skeins of Cape cormorants streaming to their fishing grounds through the fog hanging over the Namibian coastline, and similarly vast numbers of Socotra cormorants in seemingly never-ending strings skimming just above the shallow swells in the Gulf of Oman.

The Cape cormorant breeds in great colonies off the south-western African coastline but the Socotra cormorant is a non-breeding visitor to the Horn of Africa, its breeding islands located off the Arabian coastline.

No cormorant is truly pelagic. Those associated with the marine environment seldom stray far from the coast; others mix a coastal and freshwater way of life, while yet others exclusively inhabit inland waters. This is another ancient group of birds that date back some 50 million years, and by 40 million years before present they were such that we would instantly recognise them as cormorants today.

All adults of the cormorant species associated with Africa are very dark in colour, with the exception of the white-breasted form of the great cormorant. They have long, hook-tipped bills and well-webbed feet. Because the legs are set well back on the body they have an erect stance on land. When they swim their bodies lie low in the water.

The name cormorant is derived from an English corruption of the Latin for sea-crow, *corvus marinus*. Cormorants hunt their fish prey underwater, where they move with amazing speed and manoeuvrability, propelling themselves with the webbed feet and to a lesser extent with partially opened wings. After swimming and fishing, cormorants stand with wings widespread, apparently because their plumage is not fully waterproof and they need to dry off fully to remain warm. Opening the wings probably serves a thermoregulatory function, as does gular fluttering, albeit for the opposite purpose.

The common species of African fresh waters is the small long-tailed cormorant. The white-breasted form of the great cormorant is also common but usually avoids smaller water bodies. Two species, the crowned and bank cormorants, occur in low numbers along the extreme south-western coastline of the continent. Concern has been expressed about their conservation status: surviving bank cormorants number only in the low thousands, and crowned cormorants possibly fewer than 1 500 birds. ■ SEE PAGE 155.

Darters or anhingas

FAMILY ANHINGIDAE

Some taxonomists place the anhingas, or snake-birds, in the same family as the cormorants. Others believe there are sufficient differences to justify their separation. The African darter, the only member of this family occurring in Africa, is a bird of fresh waters although on occasion it is seen in tidal estuaries and coastal saline lagoons. Anhingas often swim with the body fully submerged and the long, thin neck sticking out above the surface, hence the alternative name of snake-bird.

The anhinga's bill is long, slender and pointed, lacking the hooked tip of a cormorant's bill. Also unlike cormorants, the anhinga will frequently soar when air thermals provide enough lift, and it may circle high above the colony or roost.

Unlike cormorants which grasp their prey in the bill, anhingas impale their quarry with the bill, then surface and toss the fish into the air, catching it and swallowing it. Fishing is undertaken under the water, with the large webbed feet providing propulsion. Although fish provide the bulk of their diet, they also take crabs, frogs and aquatic insects on occasion.

Anhingas nest in colonies, usually small but occasionally numbering several thousand, often in the company of cormorants, herons and ibises. They construct substantial structures in trees above water. ■ SEE PAGE 155.

Above: *The white-throated dipper is confined to fast-flowing hill streams in the Atlas Mountains.*
Left: *Anhinga at rest.*
Far left: *The darter, or anhinga, nests in colonies, either in trees, bushes or reed-beds, and often in the company of herons and cormorants.*

Dippers

FAMILY CINCLIDAE

Of the five recognised dipper species worldwide only one, the white-throated dipper, has a toehold in the north-flowing streams of the Atlas Mountains in north-western Africa. These strange birds are the only truly aquatic passerines. They have evolved a unique foraging style and are restricted to clear, fast-flowing streams where they plunge underwater, half swimming and half walking on the bottom. Here they pick up insect larvae as well as other small organisms including tiny fish. They also forage along the banks and in riverside vegetation. The wings are used as an aid when moving under water.

Their relationship to other passerines is cause for some debate. Many taxonomists relate them to thrushes whereas others favour a link with the wrens, in part on the basis of their similar large domed nests, which are constructed from plant material and have a side entrance. On leaving the nest the four or five young almost immediately take up the aquatic life style of their parents.

We have delighted in their behaviour in England and Austria, watching them foraging in the clear water and flying rapidly upstream, uttering their harsh alarm calls. ■ SEE PAGE 163.

Divers or loons

FAMILY GAVIIDAE

The divers, or loons as they are known in North America, are a small group of rather primitive waterbirds that can date their form to probably more than 100 million years before present. There are four living species, of which three are seasonally associated with the coastal waters of north-western Africa.

Divers are large, sleek and attractive birds that are highly adapted to their watery world, but this high level of adaptation has its drawbacks. The large webbed feet are located very far back on the body – ideal for swimming but making for very clumsy movement on land. In fact these are the only birds that have their legs encased in the body, and they are among a select few whose bones are hard and heavy, unlike those of most birds which are light and have large internal spaces. They are known to be able to dive to great depths but on relaxed fishing dives they seldom stay submerged for more than 45 seconds. Their small wing surface, one of the lowest in proportion to mass in the world of flying birds, would seem to indicate poor powers of flight but in fact once airborne they are strong and accomplished fliers. They need to take off from water, flailing with feet and wings for 50 m and more before rising from the surface.

In their coastal and offshore wintering grounds, the red- and black-throated divers enter the Mediterranean, but along the African coastline they are generally scarce. The great northern diver is a vagrant to African waters. ■ SEE PAGE 155.

Far left: *Lesser flamingos taking flight.*
Left: *Lesser flamingo.*
Below: *The African finfoot is a highly secretive and seldom seen bird of inland waters.*
(Photo: Richard du Toit)

Finfoots

FAMILY HELIORNITHIDAE

These secretive and highly aquatic birds have certain affinities to the rails and coots but also to the grebes, to which they are not related. Africa has a single species, one occurs in South and Central America and a third in Southeast Asia. These solitary birds favour quietly flowing streams and rivers with low, overhanging vegetation. They are lobe-footed and swim low in the water, with the peculiar habit of pumping the neck and head backwards and forwards with each leg stroke. Although most foraging for their mainly invertebrate prey occurs in the water, they also walk along the banks gleaning insects, snails and the like from the leaves and ground. When this unusual bird is alarmed it may submerge the body, leaving only the head and neck above the surface, or patter over the surface in the manner of a coot. Its secretive nature means, of course, that we know very little about certain aspects of its biology, for instance reproduction. The nest is made of twigs and the bowl is lined with fine plant material. ■ SEE PAGE 157.

Lake Nakuru in the Great Rift Valley of Kenya offers the most easily accessible flamingo spectacular in Africa. Even the tourist masses cannot detract from one of the greatest avian sights on earth.

Flamingos

FAMILY PHOENICOPTERIDAE

The ancestry of these long-legged and long-necked water-birds with their unique appearance and behaviour still puzzles ornithologists but some believe they form the evolutionary link between the herons, storks and ibises (*Ciconiiformes*) and the ducks and geese (*Anseriformes*). For example, flamingos have proteins similar in structure to those of the herons and storks, but unlike those of the ducks and geese. However, flamingo behaviour closely resembles that of geese, and flamingos share a number of parasite genera with the ducks and geese.

Flamingos are highly specialised filter-feeders. They feed and breed in huge numbers along the chain of alkaline lakes in the Great Rift Valley. These enormous flamingo concentrations are one of the natural wonders of Africa. The lesser flamingo dominates the scene, with not infrequently more than a million being present at one location. On Lake Nakuru, one early mid-year morning, we sat watching what we estimat-

ed to be more than 1,5 million of these pink and white beauties. Not only was the shoreline rimmed by them but the entire lake was a near contiguous mass of swimming birds. It is well known that both species of flamingo can swim well while feeding. On Lake Bogoria we saw a similar sight two years later, but the vents continuously belching hot steam gave it a somewhat surrealistic feel.

Flamingos feed with their heads held upside down, looking backwards between their legs. The two halves of the bill are held together, the lower mandible fitting neatly into the upper, then the head is swung from side to side. At the same time the tongue is moved back and forth thus pumping water and silt in and out. In this way tiny organisms are caught in platelet-like filters that lie inside the bill. The greater flamingo usually gathers food by completely submerging the head and filtering the mud as well as water. It also feeds on larger organisms where these are available. The lesser flamingo usually filters

Above: *A great crested grebe on its floating nest with a chick under the protective shelter of its parent's wing.*
(Photo: RPB Erasmus)
Left: *Sub-adult greater flamingos lack the pink plumage of the adults.*

only surface water, seeking the blue-green algae which are the most important component of its diet.

Numbers and movements of both flamingo species are poorly understood. There are believed to be as many as 6 million lesser flamingos, and apart from small numbers on the Yemeni Gulf of Aden they are entirely restricted to the African continent. The bulk of the population is centred on the alkaline lakes of East Africa, with one population located in and dispersing from Mauritania. Etosha Pan in northern Namibia is the principal breeding centre in southern Africa. The largest breeding concentration ever recorded was of an estimated 1,2 million pairs on Lake Makgadi in southern Kenya in 1962. A breeding colony of half a million use approximately 15 000 t of mud to build their nest mounds — a seemingly useless but nevertheless interesting calculation. These nest mounds, up to 40 cm high, are essential to the survival of the chick in its first few days of life, as the temperature in the nest cup is considerably lower than that of the surrounding mud. When a few days old the chicks of both flamingo species leave the nest mounds and form enormous "herds" constantly attended by a few adults. This is particularly noticeable in the numerically superior lesser flamingos. On such breeding grounds as Etosha and Natron the rapidly drying shallow waters force these herds to trek to permanent water which may, amazingly, be as much as 50 km away. There appear to be some exchanges between the eastern and southern African flamingo populations but details are lacking. Both species move mainly at night between feeding grounds. Their low honking calls betray their passage. Greater flamingos are not restricted to Africa but have breeding populations in Europe, Asia and the Americas. ■ SEE PAGE 156.

Grebes

FAMILY PODICIPEDIDAE

As is usual with bird taxonomy, there is some doubt as to the exact number of grebe species in the world, with a range of 17 to 22. Only five are recorded in continental Africa and none is exclusive to the continent. A further two species are endemic to Madagascar: the Alaotra grebe (*Tachybaptus rufolavatus*) and the Madagascar grebe (*Tachybaptus pelzelnii*). Three species are African residents; the red-necked and Slavonian grebes are vagrants to the northern fringes of the continent.

The smallest African representative of this family is the little grebe, or dabchick, weighing on average a meagre 150 g. The "middle-weight" black-necked grebe measures up to 450 g but usually less, and the largest, the great crested grebe, in prime condition tips the scale at up to 900 g.

Grebes are all aquatic. Overall plumage, particularly in the smaller species, is down-like and there is no obvious tail, which gives them the appearance of young birds. Their wings are proportionally small and narrow but these birds are nevertheless strong fliers. The legs are set well back on the body, causing grebes to move with difficulty on land. Their fairly long toes are well lobed as opposed to being fully webbed, but they are strong swimmers and divers. As they lack a full tail the feet act as rudders when they are flying. Another characteristic that grebes have in common is that the flight feathers are simultaneously moulted, thereby for a short period making them flightless and restricting them to the water or shoreline.

The commonest and most widespread of African grebes is without doubt the dabchick. This small waterbird occurs throughout much of sub-Saharan Africa and along the Mediterranean seaboard. Even at oases and seasonal water bodies in desert areas such as the Sahara and Namib it is not unusual to hear its high-pitched trilling call. We have even encountered one individual swimming, seemingly contentedly, in a drinking trough in the Kalahari Gemsbok National Park of South Africa.

Grebes are usually encountered singly, in pairs or small flocks, but outside the breeding season they may gather in flocks numbering in the hundreds and even thousands, frequently mixing freely with large gatherings of black-necked grebes. Although these great flocks are usually associated with large, permanent water bodies, it is not unusual to encounter them on seasonal pans. The dam that lies just above our home village in the Karoo is now, in winter, host to some 200 dabchicks and 500 black-necked grebes but when the breeding season arrives they will disperse, leaving perhaps two or three pairs of dabchicks to nest.

It has been recorded that dabchicks may benefit from the feeding activity of ducks disturbing the bottom mud. We have watched them diving among the legs of greater flamingos and following in close attendance a variety of diving and dabbling duck species.

Like all other grebes, the African residents build floating, matted nests of water weed which are attached to reed stems or aquatic grasses to prevent them drifting. Their courtship displays are all elaborate but that of the great crested grebe perhaps captures the imagination best. In Africa the great crested grebe, unlike those in Europe, retains its breeding plumage throughout the year, and its courtship rituals are not as refined but nevertheless impressive. These displays include the so-called "penguin dance" during which the pair raise themselves from the water with breasts touching, vigorously paddling with the feet and swinging their heads from side to side.

Great crested grebes have a discontinuous distribution in Africa, with the European race breeding in Tunisia in small numbers. There are population concentrations in South Africa, the Ethiopian Highlands and to a lesser extent in East Africa, where populations are said to have been decimated by the increasing use of nylon gill-nets in the principal freshwater fishing lakes.

The major threat to European populations in the mid-19th century was the demand for their skins in the world of high fashion, where they were converted into ladies' hats. Hunting pressure was so severe that by 1860 the entire population of the United Kingdom was believed to number no more than 42 pairs. To what extent hunting affected African populations, if at all, is unknown.

This, by far the largest of the three African grebes, is the "dandy" with large, prominent ear-tufts, sometimes referred to as tippets, which when spread make the head appear much larger than it is. ■ SEE PAGE 155.

The little grebe, or dabchick, is Africa's smallest grebe species.

Hamerkop

FAMILY SCOPIDAE

The hamerkop, or hammerhead, is a near African endemic, spilling over into south-western Arabia where it is closely associated with streams and wadis in the western highlands. It occurs virtually throughout sub-Saharan Africa and is common, in part because African tribal traditions have imbued the hamerkop with numerous magical powers.

The only member of its unique family, the hamerkop has uniform brown plumage, a crested head and a large, strongly flattened bill. It is one of those species, and many there seem to be, about whom taxonomists love to disagree as to which other species are its closest relatives.

This rather strange bird builds one of the largest nests known: a great domed structure with a narrow, round entrance hole. Usually situated on a cliff ledge or in a tree fork, the nest may reach 50 kg and more – an incredible feat for a bird of less than half a kilogram! The bulk of the structure is constructed with twigs and sticks but it is frequently decorated with items such as bones, wool and human debris. In the wooded grounds of a rural hospital where we lived for two years one nest was decorated with dirty bandages, drip tubes, syringes and the like – not a good advertisement for the waste disposal programme of the hospital! The entrance tunnel and nest chamber are plastered with mud as the final stage of nest building.

Nests may be used by adult birds for roosting throughout the year. Hamerkop nests are frequently taken over by bee swarms, barn owls and others. We know of one such nest that was regularly used as a retreat by a spitting cobra (*Naja nigricollis*) on the South African western escarpment.

Hamerkops are associated with inland waters and feed by wading in the shallows, catching frogs, tadpoles, small fish and aquatic insects. They will also scavenge around fishing villages. Near lakes George and Mburo in Uganda we have watched hamerkops picking up fish offal among native huts in the company of marabou storks and black kites. They will also catch prey while flying slowly over water, picking animals from near the surface with the bill. ■ SEE PAGE 156.

Above and left: *The hamerkop constructs a huge dome-shaped nest.*
Far left: *The hamerkop takes its name, which translates as "hammer-head", from the Afrikaans language.*

Above: *Cattle egret in breeding plumage, with chestnut crown and breast patch.*
Left: *The great egret, on occasion called the great white heron, is Africa's largest white heron.*

Herons, egrets and bitterns

FAMILY ARDEIDAE

The African continent is home to 22 of the world's approximately 65 species of herons, egrets and bitterns. A few are endemic but many have wider distributions. Species such as the striated heron, black-crowned night heron and cattle egret have almost global distributions. In contrast the seldom seen white-crested bittern, or tiger heron, is largely restricted to Africa's equatorial rainforests, and the slaty egret has a very restricted range centred on the Okavango Delta and the swamps of Zambia. Although many species are African residents, eight of these also have their ranks swelled by influxes of non-breeding Palaearctic migrants. The Madagascar pond heron is a non-breeding migrant to eastern and central Africa.

The members of this family are long-legged and long-necked, carry spear-like bills and have broad, rounded wings and comparatively short tails. In this family the powder-downs are particularly well developed. This strange type of feather, found in a number of bird families, is never moulted and continues to grow from the base throughout the bird's life. These powder-downs are located in paired patches on the breast, rump and flanks. They fray continuously at the tips into a powder that the bird rubs onto the other feathers to remove oil and slime from the plumage. This is followed by vigorous scratching after which body oil from the well-developed preen gland is applied as waterproofing.

Some species, such as the grey heron, are common and frequently seen, whereas others are secretive, skulking and seldom observed unless an effort is made to find them; into this category fall the bitterns. Some of the most commonly seen large birds belong to the family Ardeidae: cattle egrets hunting for insects and on occasion small vertebrates disturbed by cattle, buffalo or elephants; black-headed herons standing motionless on a road verge waiting for some animal to stir; grey herons stalking with great stealth in the shallows of a dam ...

Although most members of the family are diurnally active, several are also crepuscular and even hunt on moonlit nights. Both species of night heron, as their name implies, do most of their hunting in the hours of darkness, as do the little and dwarf bitterns. Western reef egrets seem to do much of their hunting by day but we have observed several of them feeding on a coral reef against the backdrop of a near full moon on the east coast of Unguja Island.

The black heron has evolved a unique hunting method, referred to as "canopy feeding". The wings are held open and forward, with the tips of the primaries touching the water, forming a completely closed "umbrella"; fish seen within this shadow are then caught, or at least attempts are made to do so. This umbrella pose is held only for short periods. Impressive it is when a single bird is hunting in this fashion, but on occasion larger numbers — sometimes more than 50 — take part and offer a most beautiful sight, almost like a host of black butterflies hovering and wing-flicking just above the surface.

On perusing our field notes and seeking bird anecdotes for this book we came across an entry we made for the Shingwedzi River in the Kruger National Park some years ago. The river was in spate and the road causeway was closed, so we parked nearby. Within sight was a solitary goliath heron, the largest of the family, standing patiently waiting for passing fish. On the causeway stood a grey heron also intent on prey; behind us an elephant bull was accompanied by several cattle egrets "in waiting", while a solitary little egret stood preening on a log close to the water's edge. Just below us in typically stretched-out horizontal fishing position lurked a striated heron, or green-backed heron as it is known over much of its African range, ready to strike at small fish coming within range. It was successful twice during the time we were there!

Not infrequently I see a black-headed heron hunting diurnal Karoo bush rats (*Otomys unisulcatus*) in front of my office window. Some years ago we studied the diet of these herons by collecting the regurgitated pellets from under nesting trees and analysing the contents. These largely dryland hunting herons were eating a diet dominated by small rodents, especially vlei rats, and insects, as well as smaller quantities of lizards, snakes and frogs. Fish remains appeared only twice.

Surprising was the regular occurrence of golden mole remains — as these insectivores forage just below the soil surface we can only presume that they were speared by the herons as they disturbed the soil. Birds also featured strongly, and as we know these herons will hunt by moonlight we believe that many had been taken at night from roost sites.

The grey heron is more water-dependent for hunting but it also frequently takes dryland species. In one small sample we even found the remains of young tortoises.

Many herons are slow, deliberate hunters, patiently waiting for prey, but others tend to be more active and direct hunters. We have watched western reef egrets running rapid zigzags in mangrove shallows to snatch small fish and crabs. On mud-flats on the Red Sea coast we have watched several presumably migratory squacco herons waiting in horizontal posture for fiddler crabs to emerge from their burrows. In a mangrove swamp on the Gulf of Oman we have observed Indian pond herons (*Ardeola grayii*) feeding in exactly the same manner. This pond heron has not been recorded in Africa. ■ SEE PAGE 155.

Far left: *The western reef egret, considered by some to be a form of the little egret, is a coastal heron that seldom ventures inland. It is commonly associated with reefs and mangrove swamps.*
Left: *The squacco heron has a resident breeding population in Africa, and non-breeding migrants enter from their European and Middle Eastern range.*

Ibises and spoonbills

FAMILIES PLATALEIDAE & THRESKIORNITHIDAE

Africa has an inventory of eight ibises and two spoonbills. All are fairly large, with long legs and necks. The face is naked in most species, but the black head and neck of the sacred ibis is totally naked. Ibises have long, down-curved bills; spoonbills have straight bills that are flattened and rounded at the tip, as their name implies. Most are gregarious species, frequently foraging and roosting in flocks, and most are colonial nesters, although the two forest-associated species, the olive and spot-breasted ibises, are solitary or live in pairs.

The ancient Egyptians, some 5 000 years before present, believed that the sacred ibis represented the god Thoth, the scribe. In frescos the god is depicted with the head of this bird.

Carvings were made of this ibis and many thousands of these birds were mummified and stacked in temples and the tombs of pharaohs. No wonder that today the sacred ibis is rare in Egypt. However, it is one of the commonest and most widespread ibises in sub-Saharan Africa. Elsewhere it occurs in Iraq and Iran. Despite some taxonomic confusion it is generally held that the similar ibis in Australia is in fact the sacred ibis.

One of the ibises with the most restricted range is the wattled ibis of the Ethiopian Highlands. This distinctive bird seldom wanders below an altitude of 1 500 m. The roosts and breeding sites are commonly associated with human habitation where the birds are not molested.

The waldrapp, or northern bald ibis as it is sometimes called, is one of the world's most threatened bird species. It has its last known natural breeding populations in Morocco where less than 90 pairs survive, the remnants of at least 1 000 pairs found here as recently as the 1930s. This rather strange bird used to be much more widespread but nearly all had disappeared from Europe by the end of the 17th century. A Turkish population believed to have numbered 3 000 pairs in 1890 was extinct as a breeding species by 1989. Recent sightings in the highlands of south-western Saudi Arabia and Yemen, a poorly known region, might indicate the presence of unknown breeding colonies.

Captive breeding programmes have been very successful and reintroductions have taken place but there is one serious drawback associated with this. Outside the breeding season these birds migrate southwards. According to some authorities, migration routes need to be learned from the previous generation of birds. Reintroduced colonies may therefore require the presence of wild birds.

Another ibis with a very limited range is the southern bald ibis, but despite severe population declines in the past, its numbers appear to have stabilised.

The most widespread ibis in Africa is the glossy, with resident populations occuring wherever there is suitable habitat. Migrant glossy ibises enter the continent from Eurasia. The glossy ibis also occurs in Australia and the Americas.

Both Eurasian and African spoonbills breed in Africa but the former extends into southern Europe and eastwards into Iran. Eurasian spoonbills have resident as well as migratory populations in northern Africa, but are of course absent from the Sahara. When feeding they walk through shallow water, sweeping the bill from side to side in the water and grabbing any prey item that comes into contact with the spoon-like tip. On occasion they will dash around in the shallows chasing small fish, tadpoles and the like. They usually feed in small numbers but breeding colonies can contain hundreds of birds. Nesting colonies of both spoonbills often have other species associated with them, particularly those of the African. ■ SEE PAGE 156.

Far left: *The strange, bare-necked sacred ibis was revered by the ancient Egyptians and many mummified birds have been found in temples and tombs along the Nile River.*
Left: *The Eurasian spoonbill occurs in Africa mainly as a migrant in the north, but small breeding populations are known.*
Below: *African jacanas have very long toes which enable them to walk on floating vegetation.*

Jacanas

FAMILY JACANIDAE

This is the bird with the impossibly long toes that stalks its way across the lily-pads of most of sub-Saharan Africa's suitable wetlands. Its high-stepping gait is necessary to lift the feet far enough off the ground to progress forwards without tumbling in a tangle of toes. It probes under the pads and underwater growth for the insects, snails and other organisms that make up its diet. We prefer its alternative common name, lily-trotter – at least it is explanatory and has a nice ring to it.

Although jacanas are most commonly observed singly, in pairs or small loosely foraging parties, on occasion hundreds may gather, particularly on seasonally inundated floodplains and shallow temporary pans. Sometimes stragglers turn up in

the most unusual and unexpected places. We observed one preening itself on the edge of a cattle trough along the then dry Molopo River in the Kalahari, and another at a riverbed pool without aquatic vegetation deep in the Kuiseb Canyon, which penetrates some of the harshest desert country in southern Africa. This bird was intent on catching insects, mainly flies, attracted to the damp sand at the edge of the pool. On several occasions and in different places we have seen jacanas using partly submerged hippos as "fly-traps", snapping up flies landing on their backs and heads, and also foraging in what seems to be dangerously close proximity to basking crocodiles. Apart from their amazing "walking on water" abilities they are expert swimmers and divers, this being particularly important during their annual moult as they are then flightless.

The African jacana has an interesting sex life: some birds form monogamous pairs but polyandry (a female consorted by several males) is common. The nest is a flimsy, wet platform of bits of aquatic plants, usually placed on floating vegetation. There are records of nests and eggs being moved when threatened by storms or rising water, but many clutches are probably lost. The male does the brooding. ■ SEE PAGE 157.

Top: *Black oystercatchers are a south-western African coastal endemic. The total population has been estimated to number less than 5 000 birds.*
Above: *European oystercatchers are non-breeding migrants to the coastal margins of the African continent.*

ENDANGERED SPECIES

Within the Afrotropical Region 172 bird species are deemed to be threatened or seriously threatened, and a further 93 near-threatened. Of the 172 threatened species no less than 16% are classified as endangered, meaning their continued survival is considered unlikely if the causal factors continue to operate – to put it bluntly, they will probably become extinct. Many species are threatened simply because of their restricted range, others because of habitat loss, a few because of hunting. The environmental situation in Africa and its associated islands is generally poor and progressively deteriorating, but the situation on the islands is grave indeed. In fact, 44% of all threatened African bird taxa live on oceanic islands.

A high 65% of the threatened African mainland bird species live in forests, or are forest-dependent. The most critical centres of forest bird diversity and endemism are in Cameroon, the Albertine Rift of eastern Democratic Republic of Congo, and eastern Kenya-Tanzania. It has been suggested that these three areas form the main forest refugia in Africa, where forests are believed to have survived throughout the driest periods of the Pleistocene Era. Devastation of the Upper Guinea forests has resulted in formerly common bird species with much wider ranges having become rare and threatened.

Oystercatchers

FAMILY HAEMATOPODIDAE

Oystercatchers are characterised by a long, pointed and orange-red bill, all-black or black and white plumage and reddish legs with partly webbed feet. They are specialised feeders, using the vertically flattened bill in much the same way as a chisel to prise limpets from rocks and open bivalve shells. Anyone who has tried to remove limpets from rocks must have respect for the skill and power brought to bear by these birds. They also feed on a wide variety of other marine molluscs and small crabs, and use the bill to probe for worms in mud.

Worldwide between six and eleven oystercatcher species are recognised. One of the three species that once inhabited the shorelines of Africa and its islands is extinct. The Canarian black oystercatcher (*Haematopus meadewaldoi*), always apparently uncommon, probably crossed the point of no return early in this century on its only known home in the eastern Canary Islands. Rumours persist of sightings but these are given little credence.

The African black oystercatcher is endemic to the south-west of the continent. The entire population (probably fewer than 5 000 individuals) is restricted to the coastline of South Africa and Namibia, with the occasional vagrant finding its way into southern Angola. Their basic scrape-nests, just above the high-water mark, are minimally lined with bits of debris and stones. One or two eggs are laid, mainly in the summer months. The chicks leave the nest soon after hatching and are closely attended by the parents. Sadly, this is yet another bird that faces threats to its long-term survival, in part because of its narrow habitat requirements and also because of greatly increased use of its nesting environment by humans for development and recreation. Pollution may also pose a threat but this aspect is poorly understood.

The European oystercatcher occurs as a non-breeding migrant around the African coastline and as a vagrant inland in East Africa. This bird is in a healthier state, entering Africa in substantial numbers from its breeding grounds in western and central Eurasia. It is little studied in its wintering grounds but in Mauritania it feeds almost exclusively on one bivalve species; elsewhere the diet is more diverse. ■ SEE PAGE 157.

Painted snipes

FAMILY ROSTRATULIDAE

This family has but two species, one in the Neotropics and one in Africa. Absent from tropical rainforests and very arid areas, the painted snipe nevertheless has a broad African range and extends eastwards through Asia into China, Japan, the Philippines and Australia.

In this family there is a reversal of sex roles. The female is the "showgirl" and the male a rather drab fellow who is attracted by her wing-spreading display (which he may also perform on occasion) and variously described "boooo" or "wuk-oo" call. As befits the male's lowly status he only utters an uninspiring squeaky call. The pair bond lasts just until the flighty lady has laid the two to five eggs; then she moves on to the next male, leaving each mate to incubate the eggs and raise the young. One cannot help but have unscientific thoughts of an avian soap opera or stories on the scandal pages of certain well-known newspapers ... ■ SEE PAGE 157.

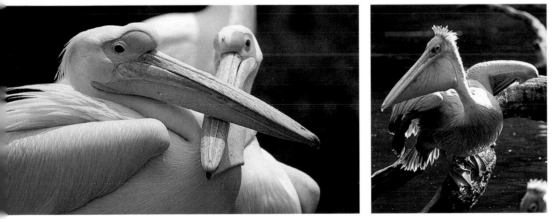

Far left: *Eastern white pelicans develop a slight swelling on the forehead during the breeding season.*
Left: *The pink-backed pelican has a wide African distribution but never reaches the numbers or densities of the eastern white pelican.*

Pelicans

FAMILY PELECANIDAE

Pelicans are of ancient origin, in fact the fossilised bones of their ancestors from the Oligocene some 40 million years ago are so similar to present-day members of this family that they carry the same genus name. The remains of species living today have been found in Pleistocene deposits.

Both eastern white (or great white) and pink-backed pelicans have extensive African distributions but the Dalmatian pelican is a rare non-breeding migrant to the Nile Delta region, with vagrants occasionally turning up in Algeria. Pink-backs occur only marginally outside Africa along the Arabian Red Sea coastline and eastwards along the Gulf of Aden. Eastern whites occur into Asia, parts of Arabia and south-eastern Europe, with birds entering North Africa when not breeding.

Pelicans are among the largest flying birds known, with the eastern white pelican male tipping the scales at up to 15 kg. Because of their large size they are dependent on air thermals when moving any distance between feeding and breeding lakes. The pink-backed pelican is considerably smaller than its eastern white relative, with adult males only averaging 6 kg. The two species also differ in feeding technique: the pink-backed is a mainly solitary fisher, whereas eastern whites are social fisherfolk. Although one may encounter single eastern whites fishing, it is more common to see groups of up to 40 individuals swimming in horseshoe formation in the shallows. Every now and then all the birds partly open their wings and in close coordination plunge their bills towards the centre of the "fishing net". When each bird has satisfied its food needs it withdraws from the fishing formation to preen and rest on the shore.

All pelicans are fish eaters, with both the eastern white and the pink-backed species favouring cichlids, this being probably more a measure of abundance than selectivity. One study of eastern whites at Lake Nakuru in Kenya indicated that the average population of 10 000 birds consumes well over 4 000 t of fish each year.

Although poorly understood, movements of the eastern white pelican appear to be triggered by fluctuations in the numbers of their food fish and their need to travel to the breeding grounds. Because eastern whites breed in relatively few but large ground-nesting colonies, disturbance, to which they are very vulnerable, is a major concern. Colonies have been known to be deserted even when large chicks are present, particularly after human intrusions. Although nearly all colonies are associated with inland waters, including some of the Great Rift Valley soda lakes where the absence of fish forces the pelicans to fly daily to fishing grounds, some are located on coastal islands. The total number of breeding pairs of eastern white pelicans in Africa is not known but believed to exceed 50 000. Larger colonies may number several thousand breeding pairs, and not infrequently they are found in association with breeding flamingos and other species, such as at Lake Natron.

Pink-backed pelicans tend to be more "area bound" than their eastern white relatives, although there is apparently some north-south-north seasonal movements in West Africa. Unlike the larger eastern white pelicans, the pink-backs nest in trees, in colonies of up to 500 pairs but usually less. We know one such nesting colony on the western shore of Lake George in Uganda; from a distance the trees appear to be covered in enormous white blossoms.

Unlike eastern white siblings which are raised in the nest with minimal rivalry or aggression, chicks of the pink-backs squabble and fight while the older and stronger individual establishes dominance over the later hatched chick. ■ SEE PAGE 155.

Left: *Long-toed lapwings show a preference for floating vegetation.* (Photo: Richard du Toit)
Above left: *Crowned lapwing chicks depend on their cryptic coloration to avoid detection by predators.*
Above right: *Brown-chested lapwing.*
Opposite page: *Three-banded plovers are common wetland foragers.* (Photo: Richard du Toit)

Plovers and lapwings
FAMILY CHARADRIIDAE

When is a plover a plover, and when is it a lapwing? It depends on which school one follows: plovers of the genus *Vanellus*, of which there are 14 species in Africa, are frequently referred to as lapwings. A further 16 species carry the "plover" tag and they fall into three genera, the *Charadrius* comprising 11 of the total. All are small to medium-sized, with shortish necks and short bills that are somewhat pigeon-like; most have long legs. The majority of species only have the three forward-pointing toes; some have a vestigial back toe. The different species occupy a wide range of habitats, often in association with water although some show a preference for drier country. Most have boldly patterned plumage and the sexes are similar in appearance. Most species, other than during the breeding season, occur in small to large flocks both when feeding and roosting. Both parents care for the young, which leave the nest scrape shortly after hatching.

The plovers, as opposed to the larger lapwings, are mostly small in size and many have white and black markings on the head and one or two dark bars on the chest. Some African residents, such as the little ringed plover, have their numbers boosted by a seasonal influx of non-breeding migrants. Near African endemics are the Kittlitz's and the three-banded plover. Commonly seen in pairs and small parties, they never form large flocks like their Palaearctic migrant cousins. These two species occur in a variety of wetlands and neighbouring habitats, but Forbes' plover, an African endemic, is closely associated with open areas such as glades and grassland within the tropical forest belt. With the onset of the breeding season they forsake these areas and move into rocky hill country.

One of the most widespread of the small plovers, the Kentish plover, ranges across Eurasia, the Americas and Africa north of the equator. Apart from the resident population, tens of thousands enter Africa as non-breeding migrants from the north, with a concentration of almost 20 000 birds at Banc d'Arguin in Mauritania.

Most species lay two eggs, occasionally three; a few sometimes lay four eggs. Most nest scrapes are simple and with little

or no lining, but some are intriguing. On several occasions we have found three-banded plover nest scrapes lined with small, flat stones in the manner of crazy-paving. One Kentish plover scrape was ringed by tiny, colourful shells but another just a few hundreds metres away was set among broken glass and ringed with glass shards.

One of our favourites is the gallant little white-fronted plover, a resident along long stretches of African coastline and penetrating deep into the interior along river courses. Like other plovers, this species forages both during the day and at night, feeding on a wide variety of insects, molluscs, crabs and other invertebrates. As is the way of some scientists, they have

calculated that on average these small waders eat over 10 g of food each day, totalling almost 4 kg in a year. Not a small quantity for a bird of less than 40 g!

The endemic chestnut-banded plover has a very patchy range, being mainly coastal in southern Africa and closely associated with the alkaline lakes near the border between Kenya and Tanzania. Among those species that enter Africa as non-breeding migrants, the Mongolian and great sand plovers concentrate on the eastern coastal margins, with only occasional stragglers getting to the western seaboard. The Eurasian dotterel overwinters mainly along the Mediterranean and partway down the Red Sea, where it does most of its feeding at night or very early in the morning.

The lapwings of Africa are dominated by mainly endemic resident species, the majority of which have fairly wide ranges. They include some of the most frequently seen, and heard, birds on the continent. The lapwings are believed to have had their origins in the African tropics from where they spread northwards and eastwards. In the south and east of the continent the sharp "tink-tink" call, oft repeated, of the blacksmith lapwing is well known. Like most lapwings the blacksmith vigorously defends eggs and chicks from any perceived danger, and on several occasions we have watched them dive-bombing elephant and buffalo. The crowned lapwing, ranging through southern and eastern Africa, is one of the few birds benefiting from some of the negative environmental impacts created by people. These include overgrazing and trampling, as well as bush clearing. Unlike the blacksmith, the crowned lapwing is seldom found near water and damp habitats. These birds become highly tolerant of human presence, frequenting golf courses, gardens and grass verges in towns and cities. Usually observed in pairs or small groups of about 10, they occasionally gather in loose associations of over 100 birds. As is common throughout the family, they vigorously defend the eggs and chicks from intruders and predators by calling loudly, circling and diving, and sometimes hitting the threat with the wings. They may also include the broken-wing display to distract predators and draw attention away from the eggs or chicks.

The attractive wattled and white-headed lapwings, with their distinctive long yellow facial wattles, differ in their choices of favoured habitat. The wattled lapwing is a generalist and although favouring moist habitats it readily occupies drier areas such as savanna, whereas the similar-looking white-headed lapwing strongly favours sandy and gravelly riverbanks. In this habitat it frequently comes into contact with basking crocodiles, and these and hippopotamuses pose a serious threat to egg clutches by crushing. The lapwings therefore vigorously try to draw potential predators away from the nest scrape or chicks, and may attack with great bravery. On hot days the adults will walk into shallow water to wet the belly feathers and return to cool off the eggs. ■ SEE PAGE 157.

Rails and their kin

FAMILY RALLIDAE

The rail family, which includes a rather mixed bag of genera and species, nearly all have one common feature: they are found in aquatic or moist environments. They range from small to medium-sized, most are dark or cryptically coloured and their family can boast more than 70 million years of development through the fossil record. Africa has a total of 28 species in no less than 11 genera. Although generally considered to be weak fliers, several species undertake long and spectacular migrations. In part, their success as a family must surely lie with their largely omnivorous diet, which includes both plant and animal food.

CRAKES & RAILS

Although similar in overall appearance, crakes have shorter, more conical bills than rails. Rails have markedly narrow, compressed bodies – an adaptation for moving quickly and easily through the dense vegetation which most of them favour. Hence the saying "as thin as a rail"! They have moderately long and stout legs, with large feet that aid walking on the wet mud and sometimes aquatic vegetation that usually dominates their favoured habitats. Black crakes (in fact a type of rail) stalk across floating lily-pads in search of small frogs and aquatic invertebrates, even carefully tilting the leaves to search their undersurface for snails and larvae. Rarely tipping the scales at more than 100 g, they barely cause the floating leaves to dip. The purple gallinule on the other hand, at more than 500 g, cannot hesitate too long in one place without experiencing that slow "sinking feeling".

Most rails and crakes are shy, skulking birds known mainly by their strident and often far-carrying calls. A certain mystery is attached to some species because they prefer to call at dusk, or even at night. Flufftails, which are small crakes, have the most haunting calls, and the buff-spotted in particular has ascended to great heights of superstition and magic. Someone hearing their calls for the first time, and not knowing their origin, can find it an unnerving experience. The source of these calls has been variously ascribed to the wail of a banshee, the call of a lady from Irish folklore whose utterances are said to foretell death, the agonising of a chameleon giving birth and the verbal outpourings of a mythical crowing crested cobra. Even giant snails and a tortoise have been "falsely accused". There are many more equally fanciful explanations for these calls, among the strangest of the avian world.

With the exception of three species of flufftail we know little about these attractive but secretive birds. All nine flufftail species, three of which are sometimes referred to as crakes, have fairly wide African distributional ranges but in most cases they are very localised. For example, the chestnut-headed flufftail is known from just a few scattered localities in the central African tropics and only in north-eastern Zambia is it considered to be fairly common.

If one excludes the endangered slender-billed flufftail (*Sarothrura watersi*) of Madagascar, the white-winged flufftail is perhaps Africa's rarest member of this group of small rails. Its disjointed distribution includes a few, probably no more than 10, small populations in South Africa, and it is believed that these all involve seasonal migrants. The largest known population at one locality, of perhaps 200 breeding pairs, is located in Ethiopia.

In most species small invertebrates make up the bulk of their diet but at least in some, and possibly all, small seeds are also eaten. Some species, again perhaps all, vigorously defend the well hidden nest and eggs. The buff-spotted flufftail, for one, has no hesitation in trying to chase off human intruders.

The Nkulengu rail is Africa's largest, and the most primitive member of this group. Unlike many other rails, this "giant" has proportionally short toes, an adaptation for walking on hard ground in the rank vegetation which it occupies along streams in tropical forest. As only one nest (which contained three eggs) has ever been found of this secretive bird, little can be said about its reproductive biology. Another species of lowland tropical forests, the grey-throated rail, is also a reproductive unknown but perhaps two of its nests have been found.

One of the best known of the world's crakes, the corncrake, is a non-breeding migrant to Africa; in fact the bulk of the Eurasian breeding population overwinters on this continent. Its summer range extends from north-western Europe eastwards to Siberia. Most corncrakes enter Africa along the principal flyways across the Strait of Gibraltar and in the east across Arabia and the Mediterranean. Although most rails are shy and retiring, Rouget's becomes very trusting of humans in parts of its exclusive Ethiopian and Eritrean range. Another unusual feature of this bird is that it has been observed diving into water and emerging some distance away, but it is not clear whether this is for foraging or perhaps for some other purpose.

Within the genus of *Porzana*, three species are "Africans". All are small with mottled underparts and are easy to recognise as rails, that is if you ever get the chance to see them. They swim and dive regularly with consummate ease when foraging for their mainly animal food. ■ SEE PAGE 157.

GALLINULES

Of the three gallinules occurring in Africa (perhaps four if modern taxonomy is strictly followed), none is restricted to it. The American purple gallinule turns up as a vagrant from the west; in south-western South Africa it occurs with some regularity albeit in very low numbers. The purple gallinule, or as it is also called the swamphen, has an extensive distribution in southern and East Africa but occurs patchily elsewhere. Beyond the continent it ranges eastwards to Australasia and even to islands in the south-west Pacific Ocean. Although favouring dense reed, sedge and bulrush beds, it not infrequently prowls the fringes and on occasion may venture onto open mud and sand flats. Its large size, showy purple-glossed plumage, short but heavy red bill and that certain touch of class ensure that it cannot be mistaken for any other member of this family. It normally forages singly or in pairs, but groups of 10 or more are not unusual. Feeding mainly on plant parts – from roots and stems to flowers and seeds – purple gallinules show great dexterity in handling food items brought to the bill by the enormously long toes. They also eat considerable quantities of animal food, ranging from insects to crabs, small fish and frogs, and on occasion bird eggs and chicks.

The lesser, or Allen's, gallinule as its name implies is considerably smaller than its cousin. Although an African species it does on occasion straggle to the strangest places, such as England, Denmark, Italy and St Helena. It frequently swims while foraging for plant and animal food. In some areas it is resident, or subject to only local movement, but overall it undertakes poorly understood migrations. ■ SEE PAGE 157.

MOORHENS & COOTS

These members of the family are the most likely to be seen, in particular the coots, which favour more open waters. The well-known common moorhen, also referred to as a common gallinule in some quarters, is actually a rail and a true internationalist. There are resident and migratory populations in Africa and elsewhere, including on suitable small wetlands surrounded by inhospitable environments such as deserts.

Lesser moorhens are sub-Saharan endemics that make use of both permanent and temporary water bodies but particularly favour the latter. The lesser moorhen differs from its larger cousin in size and in details of colour and form. It also spends far less time swimming, preferring to forage at the water's edge or on floating vegetation.

The coots are aptly described as "quarrelsome rails". The Eurasian coot occurs in the north of Africa and the red-knobbed mainly in the south and east. Although there are resident populations of both species, other large populations are migratory and the numbers involved can be impressive indeed. In the case of the northern or Eurasian form, birds moving into Africa to avoid the rigours of the northern winter number in the hundreds of thousands.

The red-knobbed coot has a large population in north-western Africa but reaches much greater concentrations in the south, with more than 30 000 on a single water body. Even on small earth farm dams in the vicinity of our home village hundreds of these slate-black birds gather in winter. Apart from the open inland waters which they favour, substantial numbers also concentrate on saline pans and estuaries. ■ SEE PAGE 157.

Left: *The purple gallinule, or swamphen, favours reed-beds.*
Above: *The black crake is a common resident of marshes and well-vegetated lake fringes but it is secretive and only observed infrequently.* (Photo: John Carlyon)
Far left: *The common moorhen has a very wide distribution extending over much of Africa, except deserts.*

Shoebill

FAMILY BALAENICIPITIDAE

The shoebill, whalehead or boatbill, as it is sometimes called, is one of Africa's strangest and rarest endemics. Somewhat prehistoric-looking, the shoebill is easy to distinguish from other large birds frequenting its limited tropical freshwater range, by its enormous bill shaped like a wooden clog. Its range largely falls within the wetlands of the Great Rift Valley but overall its distribution is patchy. The bulk of the population centres on swamps in Sudan, Uganda, Democratic Republic of Congo and Zambia, with small numbers in Tanzania, Rwanda, Central African Republic and Ethiopia. By far the highest population density is to be found in the swamps, particularly the Sudd, of southern Sudan. No accurate estimate of shoebill numbers exists, but "guesstimates" range from 1 500 to 10 000. Within its Sudanese stronghold numbers are believed to have declined dramatically in recent decades, although the civil war in that country has made it impossible to undertake any recent survey. Certainly agricultural developments and disturbance will have been a major influence, particularly as the human and cattle populations in the swamp region have increased dramatically. The shoebill is closely associated with reed and papyrus beds, frequently walking on dense floating mats of vegetation, aided by its long toes.

Shoebills hunt while walking slowly. When prey is located the bird lunges forward with its whole body, wings spread. The prey is grasped, along with a bill full of vegetation; the weed is discarded and the item swallowed. Although fish, particularly lungfish, birchirs and barbel, make up the bulk of shoebills' food they also take frogs and young terrapins. ■ SEE PAGE 156.

Right: *The shoebill, sometimes called whalehead or boatbill, has a very restricted range in the eastern African tropics.*
Far right: *The African skimmer catches small fish while on the wing, using the longer lower mandible as a "scoop".*
(Photo: John Carlyon)

Skimmers

FAMILY RYNCHOPIDAE

Skimmers are large tern-like birds. Three species are recognised by some but disputed by others who believe they are too similar to warrant separate status.

These are the only birds in which the lower mandible is longer than the upper, and this is brought into play when they are fishing. They hunt on the wing just above the water, with the lower bill just slitting the surface – hence the name skimmer. When the bill touches any fish prey the upper mandible immediately clamps down on it. The prey is flipped into the mouth and eaten in flight. Several anatomical adaptations aid this unusual feeding technique, such as the very narrow lower bill, particularly towards the tip, strong neck muscles and bony processes that help absorb the impact of hitting the prey.

The African skimmer has a wide distribution in the African tropics and occupies most suitable habitats such as rivers, lakes, coastal lagoons and swamps with areas of open water. During the breeding season they concentrate along broad, slow-moving rivers with bare or sparsely vegetated sandbanks. The one to five eggs are laid in an unadorned sand scrape, in loose colonies of up to 25 pairs. Both sexes incubate the eggs, changing frequently during hot weather. Water carried on the breast feathers is believed to be used to control egg temperature. On hatching the chick moves around the colony area. It is closely attended by one parent. ■ SEE PAGE 158.

Far left: *Black storks have a wide but localised distribution in Africa.* **Left:** *Declines in the breeding populations of the white stork in Europe reduce the numbers overwintering in Africa.* **Above:** *Marabou storks.*

Storks

FAMILY CICONIIDAE

Of the eight stork species occurring in Africa, four are exclusive to the continent, one (Abdim's stork) extends into south-western Arabia, and three (black, white and woolly-necked storks) also occur well beyond the continent. Marabou and yellow-billed storks are occasionally recorded as rare vagrants in Arabia.

All storks are characterised by their large size, long legs and necks, and long, heavy pointed bills. An interesting anatomical feature of this family is that the adults are mute; during courtship the only sound is produced by clapping the bills. All are strong fliers, with species such as the white stork migrating great distances, from Eurasia to most parts of Africa, even to its southernmost tip. The vast majority of migrating storks enter Africa along the principal flyways, across the Strait of Gibraltar and in the east via the Sinai Peninsula. Although the white stork is primarily a non-breeding migrant to Africa, there are very small resident breeding populations in the extreme south and in the north-west.

Some stork species have evolved highly specialised bills. That of the yellow-billed is extremely sensitive to touch, a clear advantage when feeding in muddy or murky water, as it allows the bird to immediately clamp the bill on any prey item touching it. The open-billed stork has a bill with the two mandibles meeting at the tip but with a distinct gap between them in the centre. It feeds primarily on water snails and freshwater mussels. Snails are held down by the tip of the upper mandible; the tip of the blade-like lower mandible is inserted to sever the columellar muscle, then the animal can be extracted. Freshwater mussels are dealt with by inserting the tip of the lower mandible between the halves of the shell near the hinge and severing the main muscles; the two halves then open.

One of the largest and most handsome of storks is the saddle-billed, an African endemic. It takes its name from the yellow saddle located at the base of the massive black and red bill. Unlike most other African storks the saddle-bill is a solitary bird. The most colonial of all, Abdim's stork, may congregate in flocks numbering thousands. Some years ago we encountered a flock of almost 2 000 Abdim's storks feeding on grassland in Hwange National Park in Zimbabwe, and more recently we watched a flock of several hundred circling over a waterpoint in the Kalahari. This pales into insignificance against reported sightings of up to 10 000 birds. Abdim's are dryland foragers, as are white storks for much of the time.

The marabou stork is probably Africa's best known member of this family. Not blessed with good looks, it has a large and heavy bill and a near-naked, black-mottled pink head and neck sprouting short, sparse "hairs" and ending in a naked, distensible pink air sac on the throat. It also has the unendearing habit of defecating on its own legs to cool down. Although marabous occasionally forage along shallow shorelines of lakes and streams, they are principally land storks, scavenging and feeding on carrion. They are master-gleaners from rubbish dumps and we have watched them selecting tasty morsels (to them!) at such sites in Kampala in Uganda and Arusha in Tanzania. We have seen them squabbling with feral dogs for possession of tasty morsels in the streets of Kampala and Entebbe. Many fishing villages along the Great Rift Valley have their regular visiting marabous, picking up scraps and then resting on the roofs of huts or in the shade of their walls. They are also regulars at large predator kills. ■ SEE PAGE 156.

Waders or shorebirds

FAMILY SCOLOPACIDAE

No less than 45 members of this family have been recorded in Africa, with the vast majority only visiting as non-breeding migrants. They include dowitchers, curlews, godwits, sandpipers, snipes, stints, knots and phalaropes – 13 different genera in all. Many are small but some are medium-sized, and all are characterised by a bill at least as long as the head, and often much longer. Most bills are straight; others are downcurved, such as in the curlews and whimbrel. Only the sanderling has no back toe, but in many species this toe is short. In most species the legs are proportionally long, as is the neck.

Many members of this family undertake long seasonal migrations. Nearly all show a strong preference for coastlines or inland freshwater shorelines, or both. Although single birds or small groups may be observed on occasion, they frequently gather in large flocks, sometimes consisting of a single species but more commonly mixed. In many species the breeding plumage differs considerably from the winter plumage, making identification difficult in some cases. Useful aids to identification are distinctive markings visible in flight, and the calls.

In the African overwintering grounds, particularly coastal ones such as Langebaan, Walvis Bay and Banc d'Arguin, literally hundreds of thousands of waders of many species gather. At Banc d'Arguin an estimated 300 000 red knots, over 30 000 sanderlings, 150 000 curlew sandpipers, 800 000 dunlins and 500 000 bar-tailed godwits, among others, may be present.

There are also those vagrant species that turn up outside their normal range to puzzle and excite observers. For example, the great knot normally leaves its Siberian breeding grounds to winter in southern Asia and Australia, but it has made landfall in Morocco. Others that pay unscheduled visits to the Darkest Continent include the long-toed stint, white-rumped sandpiper, Baird's sandpiper, pectoral sandpiper, purple sandpiper, buff-breasted sandpiper and pintail snipe.

Not all of the migrants restrict themselves to the coast. Several species penetrate deep into the interior wherever suitable habitat exists. The sanderling is relatively uncommon away from the coast but the little stint occurs in substantial numbers on muddy lake shores, floodplains and sandy riverbanks. It is the sandpipers and their kin, skittering along

beaches snatching at their small invertebrate prey, that ofttimes cause the amateur birdwatcher to falter and give up, preferring to look at the gulls over there or the oystercatcher here. In their winter plumages most are similarly coloured and marked, and only experience aids the dedicated. When they are wading on beaches or mudflats it is bad enough, but when the large flocks turn and wheel in the air in perfect unison, identification is extremely difficult.

We find the snipes and woodcock particularly exciting, in part because Ethiopian snipes utilised a small marsh on a farm where we lived for a number of years. Our attention was drawn to their presence one evening as we sat in the garden: a strange and rather ghostly "drumming" sound above us, repeated over and over again. This comes from a snipe displaying over its nesting territory; the noise is produced by the bird stooping towards the ground with the fanned outer tail feathers vibrating in the rushing air. Snipes, usually flushed singly or in pairs, rarely fly far but go rapidly to ground.

The common snipe, a non-breeding migrant to Africa from the north, usually gathers in small to large parties, occasionally hundreds, but does not feed in tight association as do many other waders. Snipes, like godwits, have very long bills for probing soft mud and wet soil for such titbits as worms, molluscs and insect larvae. The Eurasian woodcock, a close ally of the snipes, is only present in far north-western Africa. With its long bill and flexible upper mandible tip it is able to grasp its mainly earthworm prey even when it is fully sheathed with earth.

The phalaropes are a group of three smallish waders that stand out from the rest because of their feeding behaviour. The most aquatic members of the family, they land easily on the water, swim readily and perform spinning and dipping actions. Their swimming in small circles, often at considerable speed, causes small edible organisms to rise to the surface where they can be picked off with the bill. Phalaropes will also "upend" in the manner of dabbling ducks. None breeds in Africa but on the breeding grounds the female has much brighter plumage than the male, which incubates the eggs and broods the chicks. As an adaptation to their highly aquatic way of life they have lobed webbing on the toes like that of the grebes, to which they are not related.

Wilson's phalarope is a vagrant to Africa from the Americas, but the red-necked and red phalaropes occur in considerable numbers in some areas. The red-necked phalarope is found on both inland and oceanic waters, but the red phalarope is strictly marine in its overwintering areas. Normally they cluster in small groups, but flocks numbering several thousand have been counted. ■ SEE PAGE 158.

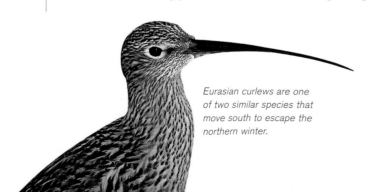

Eurasian curlews are one of two similar species that move south to escape the northern winter.

Right: *The greenshank in winter plumage.*
(Photo: Richard du Toit)
Far right: *The common sandpiper is a non-breeding migrant to Africa.*
(Photo: Richard du Toit)

Far left: *The white-fronted goose, like all true northern geese that visit Africa, very rarely penetrates to the south of the Sahara.*
Left *Male ruff in full breeding plumage; however, this species does not nest on the continent.*

Waterfowl

FAMILY ANATIDAE

Swans, geese and ducks share a number of characteristics which distinguish them from all other birds: they are relatively long-necked, usually have extensive webbing between the toes, have flattened, blunt-tipped bills (except in the mergansers, or sawbills) and the downy, well-developed young leave the nest soon after hatching.

The earliest records of waterfowl in Africa are the depictions of flocks of geese that grace the tombs and temples of ancient Egypt. Unknown artists recorded these birds as part of the bounty offered up by the life-giving waters and silts of the Nile, the mighty river that rises in the heartland of Africa.

Although the waterfowl all belong to one family, they can be divided into a number of tribes. In Africa these include the whistling ducks (Dendrocygninae), sometimes accorded full family status (Dendrocygnidae), which are characterised by their upright stance and distinctive whistling calls. The swans and true geese (Anserini) pair for life and are usually large. The shelducks (Tadorninae) lie between the true geese and the ducks in size, have distinctive and attractive plumage and replace the true geese in the southern hemisphere. The surface-feeding ducks (Anatinae), or dabblers, are well represented in Africa, both as residents and migrants. Diving ducks (Oxyurinae and Merginae) habitually dive to obtain food and

escape enemies, they tend to have rounded bodies and lie low in the water, and their large, webbed feet are set far back, making them ungainly on land.

Ducks of the northern hemisphere are sexually dimorphic: males and females have different plumages for much of the year. But in the southern hemisphere male and female ducks closely resemble each other throughout the year. During the annual moult the flight feathers are shed simultaneously and it takes three to four weeks for the new set to grow. For this period the ducks are unable to fly.

Most species are solitary nesters, with nests hidden among waterside vegetation, although a few nest in tree holes, in underground burrows or on cliff ledges. In a few species, most notably stiff-tails and pochards, there is a tendency for females to "play cuckoo" and lay eggs in the nests of other anatids.

The waterfowl of Africa can be divided into those that are endemic to the continent, those that migrate from Eurasia into Africa (whether to breed or escape the northern winter), and those that are vagrants and only rarely put in an appearance. Of the seven recognised subfamilies of waterfowl, six occur in Africa; that is 55 species out of 145. Of these, 15 are endemic; some 27 species breed in Africa's wetlands and the remainder are non-breeding migrants or vagrants.

With its great diversity of wetlands, including rivers, lakes, inland deltas, marshes and swamps, as well as its rich and productive coastline, Africa offers much for waterfowl, either resident or transitory. The lakes of Awasa, Basaaka, Koka, Tendaho and Arakit in Ethiopia are known to be important refuges for ducks and geese, and the floodplains of the Niger River in Mali attract large numbers of northern hemisphere waterfowl, providing productive feeding grounds during the cold northern winter. Coastal lagoons in Morocco are attractive wintering grounds, as is the Djoudj basin within which is located the delta of the Senegal River. Here as many as a quarter of a million ducks and geese may be present at certain times of the year, as well as many thousands of wading birds.

In South Africa, where the continent's waterfowl have been more thoroughly studied than elsewhere, most species are residents or local migrants, with very few Palaearctic species reaching the continent's southernmost point. In this country human interference has actually caused an increase in the numbers and range of certain duck species, to the extent that at least two, the Egyptian and spur-winged geese (neither are in fact true geese), are considered to be major pests in certain commercial croplands. This increase can be mainly ascribed to the many major impoundments, as well as dams on farmland, which have been constructed since the 1950s, thus creating favourable habitats for several species.

In Tunisia, in North Africa, Lake Kelbia is an important location for both breeding residents and seasonal visitors. A number of very important wetlands, heavily utilised by waterfowl, are located in Zambia, including the Kafue Flats which extend over some 6 000 km² and the Bangweulu Swamps which cover a similar surface area. Although both wetlands have seen major poaching of their larger mammal populations, waterfowl remain relatively untouched. Other major waterfowl populations occur on the soda lakes of the Great Rift Valley, such as Bogoria and Nakuru, in the great inland delta of the Okavango River and the vast expanses of the seasonally flooded Makgadikgadi salt pans in Botswana. There are permanent and seasonally inundated swamplands in Tanzania, which are poorly known and seldom visited by scientists.

Africa is a prime place for waterfowl, whether it be a flight of 10 yellow-billed ducks rising from a temporary pan in the arid interior of South Africa, a skein of spur-winged geese clattering to their roosting site at sunset, a flock of 10 000 northern pintails cruising in loose feeding flotillas in the Senegal Delta, or a pair of Hartlaub's ducks preening themselves on the bank of a rainforest stream with the pant-hoots of chimpanzees as a vocal backdrop.

SWANS

Three species of swan, the mute, whooper and Bewick's, have been recorded in Africa but the latter two only as occasional visitors and vagrants. These are all very large, white-plumaged birds, with long, slender necks, loud honking calls and a special elegance all of their own.

Both whooper and Bewick's swans fly into North Africa to escape the harsh winters which blanket their northern European and Asian summer breeding grounds with snow. Bewick's swan breeds on the tundra of northern Russia and Siberia, from the Kola Peninsula to east of the Lena River — one of the least forgiving regions on earth. Virtually nothing is known of the African haunts and behaviour of this and the

Above left: *The mute swan is arguably the most elegant waterfowl.*
Above right: *The greylag goose is a regular non-breeding migrant to north-western Africa.*
Right: *Bean geese are regular non-breeding visitors to north-western Africa but are rare vagrants elsewhere.*

whooper swan but it is generally believed that the numbers of visiting birds have declined in recent years.

The mute swan is the largest of the three great white birds. A small number of visitors overwinter in Egypt and Algeria. Egypt also has a small resident population descended from introduced birds. A similar feral group in South Africa, numbering several hundred individuals at its peak, is now extinct. The origins of this population, which was located in what is termed South Africa's lake district, are unknown but the founder stock possibly came from a ship that foundered on rocks in the vicinity of the town of Knysna. These birds flourished and occupied most of the small lakes and rivers in the area, but by the mid-1980s they had disappeared. The decline and eventual disappearance of these swans can probably be ascribed to disease. ■ SEE PAGE 156.

Below: *The yellow-billed duck is an African endemic that may congregate in large flocks when not breeding.*
Bottom: *Tufted ducks dive for much of their food.*

GEESE

Seven species of the so-called true geese migrate to North Africa to escape the harsh northern winter; they are the bean, greylag, greater white-fronted, lesser white-fronted, barnacle, brent and red-breasted geese. The first three are fairly common and regular visitors to the continent but the remaining four are vagrants.

Both the bean and greylag geese visit north-western Africa where they congregate in flocks sometimes several hundred strong. Greater white-fronted geese leave their summer breeding grounds in the Arctic Circle to spread southwards over much of Eurasia and North America, with part of the Russian population spending the winter months in the Nile Delta. At this time they are by far the most common geese in Egypt. With its distinctive white forehead, this large goose has been known for thousands of years in Egypt where it is depicted on frescos in ancient temples.

The red-breasted goose is believed to have once been a common visitor to Egypt as it features regularly in ancient frescos, but it probably never bred in that country. Its very limited breeding range is centred on the Taomyr Peninsula in northern Siberia, and today the bulk of the population overwinter close to the Black Sea in eastern Romania.

None of the true geese penetrate the Sahara, although the bean goose is occasionally recorded as a very rare vagrant to the Niger Delta in West Africa. No true geese have been recorded to breed on the African continent. However, a number of their smaller cousins do cross the Sahara. Nearly all the garganey from Europe spend the northern winter south of the Sahara in tropical Africa. Considerable numbers of common teal share these wetlands with them. Some northern migrants, such as the pintail, penetrate far into East Africa, while species such as the Eurasian wigeon and gadwall follow the duck friendly Nile Valley. There is a breeding population of ruddy shelducks in the Atlas region of north-western Africa but, remarkably, they move northwards into southern Spain during the winter months. In contrast, ruddy shelduck that breed in Asia Minor migrate down the Nile Valley to escape the rigours of winter. ■ SEE PAGE 156.

DUCKS

Four species of waterfowl endemic to the African continent confusingly have "goose" as part of their name, although they are in fact ducks and not true geese: the spur-winged, blue-winged, Egyptian and African pygmy geese. The blue-winged and Egyptian geese belong to the subfamily Tadorninae, which encompasses the shelducks, and the spur-winged and African pygmy geese are classified as true ducks, in the subfamily Anatinae. To confuse things even further, the spur-winged goose at up to 10 kg (average 5-6 kg), is Africa's largest endemic waterfowl species, and the African pygmy goose at slightly more than 0,25 kg is its smallest.

Although the other three species have very wide sub-Saharan distributional ranges, the blue-winged goose is restricted to the highlands of Ethiopia above approximately 1 800 m. Despite its small range and narrow habitat requirements, this moderately-sized (1,5-3 kg) duck is fairly common. It is indirectly protected by the religious beliefs of the local people. The greatest threat facing this and other highland species is the pressure on the environment posed by a burgeoning human population. As with many species inhabiting the high

A common duck in southern and East Africa, the red-billed teal is frequently seen associating with other waterfowl species.

plateau of Ethiopia, our knowledge of the behaviour and biology of this duck is sketchy. We do know that they occur in pairs during the breeding season, when they undertake no migration or even local movements, each pair probably holding an exclusive territory. Favouring marshes, bogs, swamps and small streams fringed with grassland, even at altitudes above 4 000 m, they graze and also consume invertebrates such as earthworms and snails. During the rainy season, which extends from July into September, the blue-wings gather in flocks of up to 100 strong and move to lower altitudes. However, it is likely that pairs retain their bond during this period and then return to their own particular territory for the next breeding season.

By contrast, the Egyptian goose is one of Africa's best known waterbirds. Once revered by the ancient Egyptians, this large goose-like duck is one of the commonest and most widely distributed of the continent's waterfowl species. It occupies a great diversity of habitats, up to altitudes of 4 000 m in the Ethiopian highlands and the mountains of East Africa. Egyptian geese can be observed along coastlines, on lakes and artificial impoundments, rivers and streams, temporary pans and seasonally flooded saline flats. During the breeding season both the male and the female are aggressive defenders of their terri-

tory, but outside this period they may congregate in flocks several hundred strong which undertake local movements and relatively short migrations. But, at least in certain areas, some pairs appear to be resident throughout the year, shunning the company of flocks even when they land to feed close to the pair's territory. Any stranger venturing across its invisible boundary is aggressively driven off.

The Egyptian goose is not particularly fussy, or selective, about where it constructs its nest and lays its clutch of five to 11 eggs on the ground, on cliff ledges, in old buildings, in trees and even on the platforms of windmills. This ability to adapt to many different situations has allowed this duck not only to hold its own but even to increase in numbers in parts of its range. A suitable nesting site is often used for several years in a row. When the authors lived on an isolated farm on the western escarpment of South Africa, one pair of Egyptian geese returned to nest in the fork of a eucalyptus tree in front of the office for five years. Another pair nested for three of those years on a rock ledge sheltered by an overhang some 80 m above the surrounding plain. We never saw these goslings leave the nest but their only means of departure was by a series of bounces down a near vertical rock wall.

Species such as the African pygmy goose and the comb duck (also known as the knob-billed duck) seek out existing tree holes for nesting, although the former is not averse to cliffs and hamerkop nests. Like the Egyptian goslings, those of the tree-nesters must "take the plunge".

The South African shelduck qualifies for the strangest choice of nest site. A southern African endemic, this duck nests in burrows excavated by the aardvark or porcupine, or more rarely the spring hare. On hatching the ducklings are escorted by the parent birds to the nearest water body, which could be up to 2 km distant. At this time the ducklings are most vulnerable to predators and raptors, but they are vigorously defended by the adults. Contrary to most birds, the female shelduck courts the male and he is the one that selects a mate. One of the most frequently seen and heard ducks in the South African interior, this shelduck usually occurs in pairs but out of the breeding season flocks of several thousand may gather during the annual moult. Their soft honking calls, often heard at night, are distinctive, and to us epitomise the high plateau country at the southern end of the African continent. ■ SEE PAGE 156.

Top left: *Ruddy shelducks breed in Africa only in the Atlas of Morocco and Algeria, but non-breeding migrants penetrate the interior along the Nile River.*
Above: *Red-crested pochards are non-breeding migrants the Mediterranean seaboard.*
Top right: *A pair of white-faced whistling ducks preening, a means of maintaining the bond between them.*
Middle right: *Common eiders include the Mediterranean in their winter range.*
Right: *Common pochards are uncommon non-breeding visitors to north-western Africa and the lower Nile Valley.*

4

Terrestrial birds

With the exception of the ostrich, all species covered in this chapter can fly, but they spend much of their time on the ground. Not only do the vast majority seek their food on the ground, most also nest on *terra firma*. Many species are fairly easy to observe, particularly those that occur in open savanna and semi-desert country: the bustards, coursers, cranes and helmeted guineafowl. Others, such as the buttonquails, the forest-dwelling francolins and the pittas, are frustrating subjects of observation for the birdwatcher! A few species have fairly close links to fresh waters, for example two of the thick-knees, or stone curlews, as well as the wattled crane and during the breeding season the two crowned cranes. Some species lead a largely solitary life but others live in family groups; still others spend much of the year in flocks of varying size. Many species are sedentary, some are subject to local seasonal movements and a few undertake substantial migrations. In this chapter we meet the world's largest living bird, the ostrich, and the heaviest flier, the kori bustard, although this distinction is sometimes accorded the great bustard, another marginal African. Sadly, this group also includes some of the continent's most seriously threatened birds. All of the cranes have suffered declines in numbers, some dramatically so, as have several bustards, certain francolins and the endangered white-breasted guineafowl of West Africa. This chapter also includes little-studied birds such as the mysterious rockfowl and the Congo peacock.

Bustards

FAMILY OTIDIDAE

Bustards are medium to large birds that are all readily identifiable as members of this family. Although occurring in Eurasia and Australasia, bustards attain their greatest diversity in Africa, which is home to 21 of the approximately 25 recognised species. There is, as usual, a taxonomic dispute which involves two to four species. In a few of the smaller bustards, known as korhaans in southern Africa, taxonomic separation is questionable. Recently the southern species known as the

Left: *Grey crowned cranes, although reduced in numbers are still common.*
Above: *Demoiselle crane head.*

black korhaan was separated, probably justifiably, into two: the black and the white-winged. In this case one cannot help but wonder why it took so long to decide to separate them.

Bustards reach their greatest diversity in southern and north-eastern Africa. A few species occur in both regions, others are exclusive to one or the other. Although the majority of species are endemic to the continent and sedentary, or at most subject to local movements in response to such factors as fire and rainfall, some are migrants. The little bustard, weighing less than 1 kg, has a tiny resident population in Morocco and a greatly reduced number of migrants entering northern Africa from Europe. This is one of Africa's most endangered bustards and is likely to disappear from its faunal list within a few years.

Except for a few of the smaller species, all bustards are under threat. Problems include drastic changes to habitats through a combination of droughts and overgrazing, but by far the most important is intensive and largely uncontrolled hunting by people in many parts. Much of this is subsistence hunting, but increasing numbers of several species are being hunted by rich individuals from, among others, the countries of the Arabian Peninsula. Having hounded the once common houbara bustard to near extinction in their own region, they have turned their attentions elsewhere. Well equipped hunting parties, which can include substantial numbers of falcons, head for Pakistan or Afghanistan, and increasingly for the Sahel and Horn of Africa. Apart from being hunted, live birds including houbaras, white-bellied and Heuglin's bustards are captured and shipped to the oil states, occasionally in frighteningly large numbers. Primitive methods of capture, rough handling and poor living conditions ensure that most soon die. Some efforts are being made to curb this trade but in our opinion not enough.

The kori bustard is generally accepted as the largest member of the family and the heaviest of all flying birds, but another giant, the great bustard, may equal or even exceed it. The great bustard occurs widely across Eurasia where its numbers have declined considerably; the small Moroccan population is so reduced that it is unlikely to recover. One cannot, however, bemoan Third World tribulations without pointing out that the last resident, breeding great bustards disappeared from England in 1833 as a result of persecution.

Another large bustard, similar in appearance to the kori, is the Arabian bustard, a strange name for a bird that occurs mainly in the Sahel belt of Africa, with only marginal representation in the south-west of the Arabian Peninsula. Other large species include Stanley's bustard, sometimes referred to as Denham's, with a distribution that sweeps across the western and central African savannas and then southwards to South Africa. Sadly, like most bustards it has been greatly reduced in number. Ludwig's bustard, a near southern African endemic, has also declined but is still a fairly common sight in some areas of the Karoo, South Africa's semi-arid interior plateau. On the extensive farmland around our home we frequently see parties of two to five Ludwig's, but occasionally as many as 20, striding slowly through the low scrub searching for insects and seeds. A red-letter day was when we observed 33 of these magnificent birds feeding on locust hoppers over a front of perhaps 1 km. Not infrequently we see them in loose association with this area's most abundant small bustard, the Karoo korhaan, which at most times consorts in pairs or threes. Other large species include the Nubian bustard of the Sahel and southern Sahara, and Heuglin's bustard with its characteristic blackfaced male. This species is largely concentrated in the Horn.

A few of the smaller species, such as the black-bellied, redcrested and the white-bellied, have wide savanna and semi-arid ranges to the south of the Sahara. If one accepts the taxonomic

Left: *Red-crested bustards have a wide but very patchy distribution in sub-Saharan Africa.*
Above left: *Black-bellied bustard.*
(Photo: John Carlyon)
Above right: *The black bustard is a near-endemic to southern Africa.*

Far left: *Karoo bustards are fairly common in the central part of their range; their frog-like croaking calls often give them away.*
Left: *The kori bustard is the tallest bustard in Africa, reaching a maximum length of 150 cm.*

thinking that the red-crested is in fact three species, each does not range very widely but the distributions of the three are nearly contiguous. However, the majority of small bustards have restricted ranges. The black bustard or korhaan, a southern African endemic, is restricted to the extreme south and south-west where the famous Cape fynbos and fringing succulent Karoo flora grow. Rüppell's bustard, or korhaan, is virtually endemic to the Namib Desert, or more correctly its fringes, and is the only resident bustard in that arid landscape. Little brown bustards are restricted to Somaliland and the Ogaden of eastern Ethiopia. The attractive blue bustard, or korhaan, inhabits the limited areas of high-lying grassland in the South African eastern interior.

All bustards are strong, powerful fliers with large, rounded wings, medium to long legs and feet that lack hind toes. They spend all of their hunting and foraging time on the ground and prefer to run or walk away from danger; if they fly it is usually only for a short distance. In many species the sexes are similar in plumage coloration and patterning, in others male and female are distinctly different. Cryptic coloration, especially in the females, offers excellent camouflage and in many cases the birds are difficult to see unless they move. However, the courtship displays of many of the males bring them to the observer's attention. The displays can be divided into ground-based and aerial, and are accompanied with distinctive calls. Every morning and evening we listen to the frog-like croaks of Karoo bustards around our home village. Black and white-winged bustards emit an ear-shattering, crowing "kraak-kraak", oft-repeated, and kori bustards utter a deep, roaring "vum-vum-vum ..." when the males are displaying.

The male courtship displays of some species are magnificent. Probably the most spectacular on *terra firma* is that of the great bustard. Males that have reached maturity, a state only achieved between the fourth and sixth year, make use of traditional lek-type areas in which they show off their finery to passing females. The tail is cocked over the back, the neck drawn into the body and the gular sac inflated; the wings are pointed downwards and backwards away from the body, with the secondaries and tertial feathers creating large white "blossoms". The kori bustard male cocks the tail and inflates the neck, causing the feathers to stand erect in a tremendous "powder puff"; the wings are extended downwards until the tips touch the ground. The smaller houbara male erects his impressive, long white and black neck plumes to envelop his head, and undertakes a zigzag or straight-line high-stepping trot.

Of the bustard males that take to the air to display, it is a toss-up as to whether the black or red-crested is the most impressive. The red-crested bustard, or korhaan, first runs along the ground and then flies vertically up to a height of about 30 m, suddenly turning on its back with feet skywards and tumbling down as though shot, with feathers puffed out, checking its precipitous descent within what seems to be a suicidal distance from the ground, only to land perfectly nonchalantly. This has quite rightly been called a "rocket flight"! Males of the black and the white-winged bustards call from a prominent point, rise to about 15 m above the ground, circle and then descend in what can best be described as a controlled tumble.

Surprisingly little is known about most of Africa's bustards, even several of those in the south where most of the continent's ornithologists and birdwatchers live. All are ground-nesters but "nest" is something of an overstatement as there may not even be a scrape and no nesting material is deliberately added. Most species lay only one or two eggs per clutch, a few occasionally lay three. The little bustard may lay the most eggs: up to six but usually fewer. Their "nests" are notoriously difficult to locate, in part because of the care the parents take in selecting the sites but also because the cryptically coloured and patterned eggs blend so well with the substrate. In some species both parents attend the chicks, in others only the female performs this duty. Hatchlings soon leave the nesting site and if predators or other danger threatens they usually squat motionless, relying on their cryptic plumage. ■ SEE PAGE 157.

The kurrichane buttonquail is more commonly heard than seen. (Photo: Duncan Butchart)

Buttonquails

FAMILY TURNICIDAE

Of the world's 16 buttonquails, only three occur in Africa, although some taxonomists recognise only two species. The black-rumped and Hottentot buttonquails are African endemics but the Kurrichane, sometimes called the little or small buttonquail, ranges widely across the continent and through Eurasia to as far as south-western China and the island of Bali. Another common name not infrequently used for these small and interesting birds is "hemipode", and even "bustard quail" has been used, apparently because button-quails and bustards occupy similar habitat and are anatomically close to each other.

All are tiny birds, with the species in Africa weighing in at less than 50 g, and they prefer to run and scuttle among the grass rather than take flight when threatened. Unlike most bird species the female buttonquails are slightly larger and more boldly patterned than the males, but nevertheless the colour of the plumage offers the birds superb camouflage in their grass-land and savanna habitats. Another deviation from the norm is that the female does the courting and is promiscuous, taking on several mates during the course of the breeding season. After each mating the female constructs the rather crude ground nest, lays her two to seven eggs, and leaves all incubation and chick care to the male. A real case of female liberation if ever there was one! Incubation may take as little as 12 days and the chick is able to run around soon after hatching. The chicks are minute at hatching, weighing little more than 1 g. Unlike quails, to whom they are not related, buttonquails are not flock-ing birds but usually occur singly or in pairs. ■ SEE PAGE 157.

Coursers and pratincoles

FAMILY GLAREOLIDAE

Of the 12 species of courser and pratincole occurring in Africa, eight are endemic to the continent. They are mostly rather small plover-like birds. Although closely related they differ greatly in their appearance and behaviour, and have been divided into two subfamilies. The coursers have long legs and short wings; generally plain in colour, most have some distinctive head or throat markings. They show a preference for open dry country, even desert, where they run rapidly, rely on their cryptic coloration and take flight only as a last resort. However, once airborne they fly rapidly. Coursers forage for insects on the ground, whereas pratincoles hunt most of their prey on the wing but also forage on the ground.

Pratincoles, whose name means meadow dwellers, are sometimes called swallow-plovers. They have short legs, long and pointed wings and forked tails. Pratincoles also differ from coursers in that they are usually associated with water, avoiding arid areas. One species favours rocks in and along rivers, another shows a preference for sandbanks. Both coursers and pratincoles are ground-nesters and two to three eggs is usual per clutch. Most coursers are diurnal or crepuscular, but one species is dominantly nocturnal. Coursers generally live in pairs, or in small flocks out of the breeding season, but the double-banded courser is usually seen singly or in pairs. All species are subject to some local and longer-distance movements but these are poorly understood. Probably the greatest influence on these movements is the cycle of wet and dry seasons, which determines the abundance of food.

The collared pratincole has the widest distributional range of any African pratincole. All pratincoles are subject to migration and local movement to a greater or lesser extent, and they may gather in large flocks numbering in the thousands. In western Zambia flocks of up to 10 000 black-winged pratincoles have been recorded in passage to their main wintering grounds further to the south. ■ SEE PAGE 158.

Above right: *Double-banded coursers, like most other courser species, are nocturnal.*
Left: *The bronze-winged courser shows a preference for wooded and bushed habitats.*
Far left and above left: *Red-winged pratincoles in the Moremi Wildlife Reserve, Botswana. Flocks may number in the hundreds.*
(Photo: Richard du Toit)

Cranes

FAMILY GRUIDAE

Of the world's 14 crane species (some authorities recognise 15), six occur in Africa; four are endemics and the demoiselle and common cranes are migrants from Eurasia. All are large, with long legs and necks, and pointed, cone-like bills. Unlike herons, cranes stretch the neck straight out in flight.

Within this family we find some of the world's most spectacular and showy birds, not only in size and adornments but also in their dances and loud, resonant and far-carrying calls which are enhanced by the long and folded trachea acting like a trumpet. Although the elaborate dances intensify during the time of courtship, dancing and prancing take place throughout the year. In flocking cranes the whole group may indulge in this. It may help to cement pair and group bonds but some maintain it is a way for the cranes to let off steam — a show of exuberance and joy. Having seen several crane species perform these dances we cannot help but favour the latter interpretation, in spite of our scientific training. The crane dances include a complex choreography of bowing, jumping with wings partly spread and calling, and each species performs a variation on the same theme.

Sadly all crane species, both in Africa and elsewhere, are threatened to some extent by habitat loss and modification, disturbance, poisoning and direct hunting. Illegal and apparently legal capture of cranes for the zoo and private menagerie trade has probably impacted on some populations. We know that hundreds of grey and black crowned cranes enter private collections in the Arabian Gulf states and are frequently offered for sale in the animal markets, or *suqs*, in the major centres. Little regard is given to international wildlife trade laws in those countries, not only in the case of cranes but also many other species.

The most abundant of Africa's endemic cranes is the grey crowned crane; also the most widely distributed, its population may number 100 000. Although threatened as a resident breeder in North Africa, the elegant demoiselle crane still occurs as a migrant from Eurasia across a broad front, with the greatest concentration in Sudan. Another non-breeding visitor from Eurasia is the common crane and although numbers entering the continent have declined, it is not threatened. The massive wattled crane, standing up to 1,2 m tall, is the continent's rarest species but sadly its numbers are poorly known and may even be considerably below the estimated 12 000. The distributional range of this magnificent bird is patchy and fragmented, primarily because of its specific habitat requirements: undisturbed wetlands. It is generally believed that the demise of the great grazing herds of game over much of the wattled crane's range caused the vegetation structure to change to the point of being no longer suitable.

The crane with the most restricted distribution is the blue, often called the paradise or Stanley's, which is a near South African endemic. And what a beautiful crane it is, with its blue-grey plumage and long streamer-like tertial feathers. Sadly,

Left: *Black crowned cranes occur throughout West Africa to northern Kenya.*
Above: *Demoiselle crane head.*

Far left: *The population of the blue crane, a near South African endemic, has declined drastically in recent years, in large part as a result of poisoning and disturbance.*
Above left: *Common cranes are non-breeding visitors to Africa.*
Above right: *The wattled crane, Africa's largest gruid, is under serious threat in the south of its range.*
Left: *Demoiselle cranes are virtually extinct as a breeding species in Africa but migrants extend down the Nile Valley to Sudan and southern Ethiopia.*

neither its beauty and elegance nor its status as South Africa's national bird was enough to prevent its rapid decline. In some areas it has disappeared completely, in others it has suffered declines of as much as 95% in less than 20 years. Even more disturbing is that this decline seems to have taken place largely unnoticed until fairly recently. More than any other African crane, the blue has adapted to the extensive tracts of cultivated lands across its range, but in many ways this has been its undoing. Until recently, and even today, deliberate and accidental poisoning on farmland is the single most important factor in this bird's decline. Blue cranes pair off during the breeding season but at other times they form flocks of 30 and up to 300 individuals, and what a spectacle this is! We shall never forget the sight of 80 of these cranes, heads into the gale-force wind, standing tightly bunched on the edge of a storm-tossed farm dam in the wheatlands of south-western Africa.

Another national bird is on the verge of extinction in its own country: the black crowned crane of Nigeria. Yet a few years ago it was considered to be abundant in the north of the country. It has also seen declines in other areas, such as Waza National Park in Cameroon, where in the 1970s an estimated 10 000 of these fine birds were present – no longer. Previously considered to be a subspecies of the grey crowned crane, it is now deemed distinct enough to warrant its own full "species tag".

Cranes are believed to pair for life and retain this bond even when flocking outside the breeding season. On their wintering grounds common cranes may congregate in flocks of up to 1 000 strong but during migration the flocks number in the low hundreds. They can fly at altitudes above 6 000 m, at speeds of 40-60 km/h or more, and maintain this for over 20 hours. Although most cranes follow the usual "narrowest point" crossings of the main flyways into and out of Africa, many make longer sea crossings. Demoiselle cranes migrate in small to large flocks but at much lower altitudes than common cranes. Africa's endemic cranes undertake some movement within their ranges, with some making longer migrations than others.

All cranes nest on the ground and most in close association with water, sometimes on trampled mats of reeds and grass surrounded by marsh, swamp or other wetland. Most species, such as the wattled, lay only one or two eggs, but on occasion the blue crane female lays three, and four have been recorded for crowned cranes. Both the male and female incubate the eggs, care for the chicks and vigorously defend them with distracting tactics, such as wing-dragging and dancing. With their slow rate of reproduction, cranes are among the longest-lived birds, 50 years not being unusual. A lifespan of 100 years has even been claimed but no documented evidence of this advanced age exists. ■ SEE PAGE 157.

Dikkops, thick-knees or stone curlews

FAMILY BURHINIDAE

The common name to use for this small but distinctive family is very much a personal choice! Dikkop is in general use in southern Africa, and a mix of all three names elsewhere. Four species occur in Africa; two are endemic and one is a near-endemic. All are medium-sized, plover-like birds with cryptically coloured plumage, large heads and eyes — a clue to their mainly nocturnal activity — and long legs with the characteristically enlarged leg joint, hence the name thick-knees. They have only three short but fairly stout forward-pointing toes which leave a distinctive track in mud and silt. Male and female closely resemble each other and may remind one of a small bustard in general appearance.

Two species occupy a range of dry habitats, whereas the other two are strongly associated with rivers, lakes and mangrove swamps. The stone curlew and Senegal thick-knee occur north of the equator in Africa, the water thick-knee in similar habitat but their ranges do not overlap. The spotted thick-knee has the greatest range, through southern and East Africa, into the Horn and westwards across the savannas to the Atlantic coast. There are also breeding populations in southern Arabia.

The thick-knees, in particular the spotted, can become very tame and are commonly found in association with humans, such as on golf courses, in large gardens and the like. Despite this they are not always easy to see, as they stand or sit in deep shade and do not flush easily. Their characteristic ascending, whistling call is one of the classic sounds of the African night.

The number that may congregate together varies from species to species, season to season and region to region, but in general they keep in pairs or small groups. Spotted thick-knees may gather in loose groups of 50 or more after the breeding season, and stone curlews form flocks on occasion numbering more than 100 birds.

All thick-knees have a broad diet, taking a host of insects and other invertebrates, small lizards, mice, birds' eggs and chicks, frogs and even some seeds. It would seem that they eat just about anything they can get their bill around. On Unguja Island, Zanzibar, where we were doing surveys, we encountered many water thick-knees at night: virtually each rain puddle — and there were many — was attended by one or two of these birds. Most appeared to be catching the numerous small frogs that were crossing the tracks and calling from the pools.

Thick-knees lay their eggs on the ground in a shallow scrape, which is lined and ringed with bits of vegetation, small stones and antelope, sheep or goat droppings. ■ SEE PAGE 158.

Above left: *Spotted dikkops, or thick-knees, are mainly crepuscular or nocturnal.*
Above: *Water dikkops, or thick-knees, occur on the edges of lagoons or rivers.*
(Photo: Richard du Toit)
Left: *The Senegal thick-knee is similar in appearance to the water thick-knee.*

Helmeted guineafowl drying off after a rain shower.

FOWL-LIKE BIRDS
ORDER GALLIFORMES

This order derives its name from *gallus*, which is Latin for "cock". In the past the term "gamebird" was often used to describe this group (and some others, notably ducks), because they were popular hunting targets, particularly species of open grassland and uplands. Only in South Africa has this had any real economic impact.

Africa is endowed with one of the world's three peacock species, 42 francolins and partridges, four quails and all six guineafowl species. The fossil record shows that their ancestors were already well established in the middle Eocene Era some 45 million years before present. This is the order in which domestic fowl and turkeys originated, and is therefore the one most closely allied to humans: where would we be without roast chicken and turkey? The helmeted guineafowl has also been kept in domesticity for centuries, and the lowly quail is farmed for harvesting.

All African forms are basically terrestrial, although many roost in trees or on cliffs. Most are sedentary but there are a few exceptions: quails, in particular, are "restless souls". Most are stocky and stoutly built and have large, heavy feet with three well-developed toes facing forward and a much shorter one behind. Although most fly readily on their short, rounded wings, many prefer to run, doing so with a good turn of speed. The males of many species, particularly the francolins, are equipped with sharp spurs on the legs, well above the foot. These spurs are used in conflict with other males of the same species and possibly also against potential predators. All have short, stout and slightly down-curved bills, which along with the powerful feet are used for digging and scratching out seeds, corms and insect food. All are ground-nesters and the chicks are precocial: they are well developed at hatching and almost immediately move away from the nest site. Within two weeks the chicks are capable of limited flight.

These birds all have grating, loud calls, far from anything that can be called musical. Yet many of the calls are so redolent of Africa that one enjoys hearing them.

Left: *Although Swainson's spurfowl is common throughout much of its range, declines have been noted in South Africa.*
Below: *Quails are tiny francolin-like gamebirds with both resident and migratory populations.*

Francolins, partridges and quails

FAMILY PHASIANIDAE

This family consists of 12 different genera in Africa, of which *Francolinus* with 36 species is the largest. Africa is the true heartland of francolins, the continent's answer to the grouse of the northern hemisphere. Only five francolins occur elsewhere, all in Arabia and Asia. Erckel's francolin, a near endemic to Ethiopia and Eritrea, has been introduced in Italy where it is relatively abundant and successful.

Although a few francolin species have fairly broad distributional ranges, many have very limited ranges, and a few are so restricted that they are threatened with extinction. Francolins range in size from small (barely tipping the scales at 200 g) to medium (over 1 kg). In most species males and females are similar in coloration and all are well camouflaged.

As a group, francolins occupy virtually all African habitats, from sea-level to 4 000 m, with the exception of true desert. Many species favour different savanna areas. The Djibouti francolin, only known to occur in the Forêt du Day which covers little over 1 400 ha, is seriously threatened and probably endangered. This primary forest has undergone massive modi-

fication and destruction over the past 20 or so years. It is hard to imagine, but the Day Forest once extended over 400 000 ha. It is difficult to assess the size of any bird population, and never more so than in forest habitats, but an estimated maximum of 5 000 Djibouti francolins survive today. At the present rate of habitat destruction this fine gamebird could be extinct before we see the new century in! Certain species with restricted ranges are, however, not under any threat, and in some cases their numbers have actually increased.

Three other species are of known conservation concern, but our knowledge of several more francolins is so poor that we have no idea of their status. One species known to be under threat is the Mount Cameroon francolin, which occurs only on the forested slopes of that western African mountain. Apart from the usual dangers posed by restricted range, hunting and habitat loss, this species faces a threat almost unique in the bird world and for a change not to be blamed on *Homo sapiens*: volcanic eruption.

A few species have relatively wide but discontinuous distributions, for example the coqui, red-wing, Shelley's, crested, scaly, double-spurred, Clapperton's and red-necked francolins.

The one francolin that we personally know better than any other is the Cape, found in a narrow belt in the fynbos, or heathland, of south-western South Africa, and along the lower

Orange River. This large bird lives in small groups, or coveys, and can become very trusting if not hunted or harassed. In suitable habitat with an abundance of food (in particular small bulbs and corms), they may reach high densities. When we lived on the western escarpment of South Africa we estimated that no less than 300 Cape francolins shared our valley. Pairs and coveys seldom mixed but one could sometimes count up to 10 coveys foraging within a few metres of each other.

Some francolins are easy to see and make no great effort to rush for cover unless obviously threatened, but others are secretive. It is often their harsh calls that give them away to the human observer. The forest and montane francolins are not too difficult to observe if one has the patience to sit quietly in glades and at forest edges. The scaly francolin is frequently seen at higher altitudes on both Mount Kenya and in the Aberdares, and less frequently the moorland francolin. In the African savanna conservation areas most frequented by tourists a number of species can be commonly seen, with the rather bantam-like crested francolin probably being the most frequent. Species that become tourist-camp regulars include the Natal, red-billed and Swainson's francolins in southern Africa, and the red-necked, yellow-necked and coqui francolins in East Africa.

What is amazing about francolins is how little we really know about most species, even those that occur in substantial numbers or have fairly wide distributions. We do know that all francolins forage on the ground but to escape the attentions of predators most roost at night in trees, bushes or occasionally on rock ledges. A few, such as the coqui and Schlegel's francolin, roost under cover, on the ground. The vast majority eat mainly plant parts, such as bulbs and corms which they dig out with feet and bill, and on occasion small quantities of invertebrates such as termites. A few, however, eat mainly invertebrates. One such is Latham's forest francolin, which inhabits equatorial forests; termites, ants, beetles, snails and other invertebrates make up 90% of its daily intake. Another species favouring large quantities of invertebrate prey is the white-throated francolin of the West African savannas. Most species have seasonal differences in favoured foods, for example Cape francolins: during summer they mostly scratch out corms, bulbs and roots but during the winter rains they eat substantial quantities of green plant material. Many francolins are also opportunistic in their feeding and we have encountered several species scratching and picking through animal dung, most notably elephant, rhino and buffalo.

In many areas several francolin species may occur close to each other, but they are usually separated by preferred habitats and possibly subtle dietary differences. In Namibia where three species of francolin commonly occur in overlapping ranges, the red-billed occupies bush alongside dry watercourses,

Above left: *The crested francolin is abundant in the eastern half of Africa.*
Above middle: *Shelley's francolin.*
Above right: *Yellow-necked spurfowl are among the more conspicuous francolins in East Africa.*
Left: *Chicks of all francolin species, including this Cape, are cryptically coloured.*

Above: *Vulturine guineafowl are the most attractive and distinctive in their family.*
Right: *The harlequin quail is a gregarious species.*

Hartlaub's the broken hill country and the Orange River francolin the grassland lying between the other two habitats.

All francolins are regular dust-bathers, and their vigorous scratching and dusting create shallow, rounded depressions. In the case of species found in areas where rock hyraxes occur, such as the Cape francolin, the mammals and birds frequently make use of the same dust-bowls – obviously at different times.

Francolins split into pairs, away from the covey, during the breeding season. Most species, if not all, vigorously defend the area around the nesting site against others of the same species. Nests are always on the ground, usually under cover, and are shallow scrapes with a greater or lesser amount of plant material used as lining. However, the only known nest of Hartlaub's francolin was located on a rock ledge, and the single Nahan's francolin nest described was 1 m above the ground in a tree fork. Are these the norm or are they the exceptions for these species? We know nothing about the clutch size, incubation or development of several species.

Known clutch sizes range from two to 12 eggs, with the norm believed to be between four and six. Apparently the female Lathan's forest francolin is usually content to lay but two eggs. It would seem that francolins are good parents, and the young remain with the adults until they are fully developed. Young francolins are able to fly short distances long before they reach adulthood, and some may at this stage be mistaken for quail.

In some literature the smaller francolin species are referred to as partridges but there are only five true partridges, in four genera, in Africa. The Barbary partridge is a North African endemic, but it has been introduced into southern Spain, the Canary Islands and apparently Sardinia. Although it is largely restricted to the coastal plain and the Atlas range, populations isolated by the Sahara are found in at least two of the great desert mountain massifs.

The sand partridge is a bird of hilly and rocky desert country with a wide range in Arabia, but in Africa it occurs only along the Red Sea Hills in Egypt and Sudan. We have not encountered this species here but know it well from the United Arab Emirates and Oman. As is the problem with common names, why lumber the poor bird with the name "sand partridge" when it seldom ventures away from dry, rock-strewn country? This bird served on many an early Arabian morning as our natural "alarm clock" – not a particularly attractive sound but certainly far better than the conventional clanging and ringing of everyday wake-up calls.

A much more widespread African species is the stone partridge, closely associated with rocky, densely bushed terrain in the broad Sahel belt, with a few apparently isolated populations in Ethiopia. In Mali it has been observed in more open, sandy country but always with access to good vegetation cover.

Only in recent years, in fact in 1991, the Udzungwa partridge came to light. This small, distinctive and rather attractive gamebird is only known to occur between an altitude of 1 250 m and 2 000 m in forest patches on the Udzungwa Mountains of south-central Tanzania. It has never been studied and we have absolutely no idea of its requirements or biology. What is really intriguing about this recently discovered bird is that it apparently has no strong affinities to any other African gamebird, but instead to the Asian hill partridges!

A small but stable introduced population of the chukar (*Alectoris chukar*) lives on Robben Island.

No true pheasant occurs naturally on the African continent but the ring-necked pheasant (*Phasianus colchicus*) was introduced into northern Morocco for sport hunting. Whether they have established feral breeding populations is unknown to us. We know of limited releases of this same pheasant in parts of South Africa but none survive today.

Three true quails, as opposed to the unrelated buttonquails of the order Gruiformes, occur in Africa, but none are endemic to the continent. They are the common, blue and harlequin quails. The common quail has by far the widest distribution: virtually throughout Africa, except in the tropical forests, and throughout Eurasia to central Siberia and India. It may reach a mass of slightly more than 100 g. The smallest, the blue quail, tips the scale at just over 40 g.

All quails resemble miniature francolins. They are difficult to flush from cover but when they do emerge it is with a rapid whirring of their short, rounded wings, only to land again about 50-100 m further on.

All three are migratory, or at least nomadic, out of the breeding season. The common quail most clearly undertakes seasonal migration, and in some years large numbers arrive at the wintering or breeding grounds. There are amazing stories about the vast migratory flocks of yesteryear. At the beginning of this century more than two million wild quail were caught each year and exported as food from Egypt alone. By 1920 the number had risen to some three million birds. Further millions were slaughtered in Asia. Obviously the populations could not sustain this massive harvesting, and numbers have never again built up to their former strength. Although hunting of these small gamebirds continues at low levels in some parts of their range, quail are now bred in captivity in several countries, both for their succulent flesh and small but tasty eggs.

All quails are opportunistic feeders, but seeds dominate the diet of common and blue quails, supplemented with a variety of invertebrates, such as beetles, grasshoppers and termites. Insects are more important as a source of food to the harlequin quail. ■ SEE PAGE 157.

Guineafowl

FAMILY NUMIDIDAE

There are now six recognised guineafowl species, in four genera. They are among the largest members of the order Galliformes. Their heads and necks are bald to a greater or lesser extent, but they usually have some type of head decoration, ranging from the horny crest of the helmeted to the "curly tops" of the crested and plumed, and the reddish "Mohican tuft" of the largest of the six, the vulturine guineafowl. Four species are associated with forest and dense woodland. All have short, rounded wings and although they can fly strongly they prefer to run. When compelled to fly, they invariably do so only for short distances. For much of the year they are social birds

Vulturine guineafowl live in flocks in dry scrub and wooded areas of East Africa.

Left: *Head and neck of a male southern ostrich – the head is very small in relation to the overall size of this mighty bird.*
Far left: *Crested guineafowl from the southern part of their range. Although commomly heard, this species is not easy to see in its forest or woodland habitat.*

living in flocks of varying size, but during the breeding season they pair off and move away from the flock. Although all are ground-feeders, gleaning their food by picking off the surface or scratching at the soil with the feet and bill, we have seen helmeted, vulturine and crested guineafowl pecking fruits from shrubs and low bushes. Along the Ewaso Ingiri River in Kenya we have watched the vulturine guineafowl jumping a few centimetres off the ground to pluck the fruits of *Salvadora*.

Certainly the best known of the six guineafowl species is the helmeted. Although absent from forest and desert, the helmeted guineafowl has an almost complete sub-Saharan distribution, with an isolated and increasingly threatened population in Morocco. Wherever there is open vegetated country one invariably finds helmeted guineafowl. Anyone who has slept close to one of their tree roosts will never forget it — off and on throughout the night their raucous cackling keeps man and beast awake.

If one takes the time to watch foraging flocks one soon notices a pecking order in the ranks, much in the way of domestic hens, but in the case of guineafowl it is the males that do most of the squabbling and arguing. Flocks will attack potential ground and aerial predators with vigour. The bulk of their food is made up of seeds, corms, small bulbs and certain insect prey when abundant. Favourites are termites and grasshoppers. Once the six to 12 eggs have hatched the chicks, or keets as they are sometimes called, stay with the parent birds for between one and three months before the family rejoins the flock as a group.

There is a small population of helmeted guineafowl in western Yemen of Arabia. It has been suggested that they descend from introduced birds but the fact that a number of primarily African birds occur in that part of the world suggests that these guineafowl constitute a natural population.

By far the least known guineafowl is the endangered white-breasted, which occurs in a few areas in the rapidly dwindling Guinean forest block of West Africa. Much more widespread in the past, it is now very restricted. Even though it is known to occur in a few conservation areas, none is properly managed; poaching and other illegal activities are rampant. Surviving numbers are not known but in parts of south-western Ivory Coast its density has been estimated at up to 12 birds per square kilometre. However, density estimates made in limited areas and extrapolated for much larger areas need to be treated with caution. One estimate places the total population of this bird at 60 000, but given its habitat and the ongoing pressures it faces the number could well be considerably lower. An interesting behavioural observation from Tai National Park in Ivory Coast is that these guineafowl rely to an unknown extent on plant parts dropped from the trees by foraging monkeys. It has been suggested that heavy hunting pressure on monkeys (a food favoured by people in that part of the world) may have had a negative impact on these birds. The black guineafowl, a resident of dense primary forest in a limited area to the north of the Congo River, is equally poorly known but we are not aware of any immediate threats to its survival.

The two *Guttera* species are distinguished by their black, curly topknots. The more widespread crested guineafowl, which has been better studied than the plumed, is a bird of forest edges, riverine forest, dense thickets and woodland. One often hears these birds, their call less grating than that of the helmeted, as they work their way through woodland trails; one sees their tracks in the mud, or one picks up the odd feather — but one seldom catches a glimpse of them. On one farm in the Soutpansberg with which we are familiar, the owner was unaware that they occurred on his property. On Unguja, Zanzibar, we once sighted a flock feeding in association with an Ader's duiker and a troop of Zanzibar red colobus monkeys.

Certainly our favourite is the vulturine guineafowl of East Africa, to us this is the aristocrat among "common" squabblers! It is the largest of the family, brilliant blue-black in colour, and restricted to arid East Africa and into the Horn of Africa. ■ SEE PAGE 157.

Ostrich

FAMILY STRUTHIONIDAE

The ostrich is the largest of all living birds, reaching a height of more than 2 m and a mass ranging from 60-80 kg in adults, with males being larger than females. With its large size and long legs and neck, the ostrich is flightless and has soft, drooping plumes, particularly on the wings and tail. The male's plumage is predominantly black and white, whereas that of the female is brown to brownish-grey. Although up to six different races have been recognised, differences lie primarily in the leg and neck coloration of the skin of the male; probably only three races are different enough to be separated. Ostriches have only two forward-pointing toes on each foot, the longer of which is armed with a heavy, blunt claw that can be used with deadly effect when attacked, or in defence of chicks. Man is not immune and several people have died from injuries inflicted by ostriches, both in the wild and in captivity.

In spite of this formidable weaponry to confront enemies, ostriches survive predation mainly by their extreme alertness and especially their speed, 50-60 km/h not being unusual. A wide range of predators hunt them, in particular leopard and cheetah. If pursued they frequently jink from to side to side to confuse the predator. Few, if any, predators can outrun a flee-ing ostrich, but they are usually taken when at water, or feeding, at rest, or when incubating eggs.

Ostriches normally occur in small flocks of up to 30, but in arid areas temporary aggregations of several hundred may gather at water, or in localised areas where recent rain has produced a green flush. We have observed some 200 ostriches feeding on a gravel plain in the central Namib Desert shortly after rain had encouraged fresh growth. Interspersed with the birds were more than 100 southern oryx and 30 Hartmann's mountain zebras – a truly magnificent sight.

Although commonly believed to be omnivores, ostriches in fact rarely take animal food. A wide variety of plant species and parts are consumed, but they can be highly selective. In arid areas they eat large quantities of succulents, thus obviating the need to drink. Where water is available they will drink regularly but this is not essential to their survival.

The ostrich, as befits its great size, lays the largest egg of any living bird. However, in relation to the bird's own size it lays one of the smallest eggs. The eggs weigh between 750 g and 1 600 g and have a capacity equivalent to about 24 chicken eggs. Although ostriches tend to form monogamous pairs, it is

THE MYSTERY OF THE CONGO PEACOCK

An avian tale that could well have been written by Agatha Christie is that of the discovery of a new species for the continent, the Congo, or African, peacock (*Afropavo congensis*). In 1913 the ornithologist James P. Chapman was exploring the Ituri Forest near Avakubi, in what is now the Democratic Republic of Congo, and noted the feather of an unfamiliar bird stuck in the hat of one of the locals. He took the feather back to the United States of America but despite great effort the mystery bird was identified only in 1936. Chapman, on a visit to the Congo Museum in Tervuren, Belgium, found two mounted specimens of peacock-like birds. They were labelled as juvenile Indian peacocks but he immediately realised that the identification was wrong. Examining the wings of the female specimen, he recognised the same brown, black-tipped feather that he had obtained 23 years earlier in the Ituri Forest.

Chapman wrote the first description of this peacock-like bird, the only representative of this distinctive group on the African continent.

A Belgian, De Mathelin de Papigny, who had lived in the Congo for many years, described a large, unfamiliar bird that he had eaten in 1930. This was almost certainly the Congo peacock.

A Belgian government employee, T. Herrlig, based at Ikela, Tshuapa, was the first person known to have kept the Congo peacock in captivity, in 1938. Several zoos, mainly in Europe and North America, now have captive breeding populations of this rare and elusive bird.

Above: *Mating pair of Masai ostrich.*
Left: *Ostrich chicks are superbly camouflaged and difficult to see while they remain still.*

not unusual for a male to mate with two or three females. The female lays three to eight eggs, depending on her age and social ranking. Other females may pass through the territories of several males, laying eggs in nest scrapes even though they have not mated with the territorial male. Large numbers of eggs in a nest scrape indicate that more than one female deposited eggs there. It is also not unusual for single eggs to be dropped at random as the hen birds move about feeding; obviously these will not be incubated. Active incubation of the eggs begins about 16 days after the first egg is laid by the major or dominant hen. Both male and female incubate the eggs, he during the hours of darkness and she during the day. Because of the staggered laying, incubation takes between 39 and 53 days, but on average 42 days.

The cryptically coloured chicks are able to follow the parent birds soon after hatching and they are closely guarded. Despite this parental care many chicks fall prey to predators. Although capable of running within a short period of hatching, chicks frequently lie flat on the ground, relying on camouflage to avoid detection. Broods from several nests may join together to form a large creche, which is accompanied and aggressively guarded by one or more adults. The largest such creche on record involved no less than 380 young birds.

Ostriches once roamed much of the African desert, semi-desert and savanna but they have been hunted to extinction over much of their North African range, and elsewhere are largely confined to game reserves. Until the middle of this century they still occurred in Arabia but the last known birds were shot in Syria and Saudi Arabia. The ostrich lineage goes back a long way and five fossil species are known, the oldest having roamed the earth between 50 and 60 million years ago. Apart from Africa, these ancient runners are known to have occurred in southern Europe and in Asia as far north as Mongolia.

Ostriches have played a role in the human history and civilisation for thousands of years. In an ostrich feather the rachis, or quill, runs exactly down the centre of the plume, there being an equal amount of "feathery material" on either side. Because of this quality the ancient Egyptians used the ostrich plume as a symbol of law and justice. Cups made from ostrich eggs have been found in Assyrian tombs that date back some 5 000 years before present, and similar artefacts from the civilisations of the ancient Egyptians, Greeks, Romans and Chinese. The body fat of ostriches was used as a drug by Roman physicians and they believed that gizzard stones could help cure certain eye diseases. It was the fine plumes of the male ostrich that most frequently appeared in heraldry during the period of the Crusades, and several centuries later they became fashionable in the Western world.

Ostrich plumes with the greatest commercial value are the 16 on each wing and the 50 to 60 tail plumes which grow in layers over the 14 true tail feathers. Less commercially valuable body plumes are used to produce such items as feather dusters. By the end of the 19th century plumes were fetching very high prices but the source of wild birds was dwindling – enter the ostrich boom. Ostrich farming with tame birds had its origin in South Africa, when several farmers succeeded in breeding and raising ostriches in captivity in about 1863. With the invention of the incubator in 1896 the numbers of ostriches farmed increased rapidly. By some estimates there were as many as one million ostriches being farmed in South Africa in 1914, but with the First World War the trade collapsed and by 1930 only some 23 500 birds remained. Until recently there had been a major revival in the fortunes of ostrich farming, but this time the principal demand was for their skins, which can be processed to make fine leather. Meat and feathers are now seen mainly as a byproduct. ■ SEE PAGE 155.

Above: *Male ostrich drinking.*
Left: *An ostrich party enjoying the lush bounty that follows rain in the Kalahari.*

Pittas

FAMILY PITTIDAE

Pittas are very short-tailed, plump ground birds that are secretive inhabitants of the forest floor within the tropics. With only two species occurring in Africa, pittas reach their greatest diversity in Southeast Asia. They have short, rounded wings but despite this the African pitta, in East Africa, undertakes seasonal migrations. It is at these times that most pittas have been collected as they become confused by lights at night. These beautiful birds are brightly plumaged with clear greens, reds and blues, and black and white markings on the head, yet despite their rainbow colours they are seldom spotted in their forest homes. When disturbed they seldom fly but scurry away into the dense undergrowth. Apparently they roost in trees and underbrush. Like broadbills, pittas are believed to be descended from primitive stock and as a group may be in evolutionary decline. Small invertebrates make up their entire diet, many of which are located by intense "listening stops" and the flicking aside of leaf litter with the fairly stout bill. When not breeding they live solitary lives, each bird occupying an exclusive foraging range. Their large domed nests, constructed with twigs and other plant material, have a side entrance, and are placed on horizontal or gently sloping branches, and at least in one species also in dense plant tangles. ■ SEE PAGE 161.

Male Masai ostrich in the Serengeti.

A male Cape rockjumper. (Photo: JJ Brooks)

Rockjumpers and rockfowl

FAMILIES TURDIDAE & PICATHARTIDAE

The bare-headed rockfowl (genus *Picathartes*) are sometimes placed with the babblers, at others with the crows, at times on their own, and on occasion they are grouped with the two rockjumper species (family Turdidae) of southern Africa, a line we have chosen.

Two species of rockfowl are known from West Africa but indications are that a possible third species may be present in the vicinity of the Kasinga Channel, linking lakes Edward and George, in western Uganda.

Little is known of these extraordinary birds, but they inhabit rainforests, especially in areas with caves. Their food is mainly invertebrates. They are medium-sized, with crow-like bills, long legs, bare heads and black or grey and white plumage. Their tails are long but they manage only weak flight on their rounded wings. Although gregarious, they are rather quiet. The nest is a massive lined cup of mud, usually cemented to rock walls in caves. Two eggs are laid.

The two rockjumper species (genus *Chaetops*) are endemic to southern Africa. Small passerines belonging to the thrush family, they live in pairs or small parties on rocky mountain slopes. The males are handsome with black and rufous or orange plumage; the females much duller. ■ SEE PAGE 164.

5

Raptors of the day and night

Africa and its associated islands boast approximately a third of the world's diurnal birds of prey and some 40 owl species. In this assemblage many are endemic, or near endemic, and many others enter the continent as seasonal migrants. These include the avian world's finest and most efficient hunters, with anatomical features such as powerful feet and long, sharp talons, and hooked bills adapted for tearing prey. Raptors are variously treated with veneration and with hatred by man, as both friends and enemies. Much has been written about them, not least by that master of literature William Shakespeare. Here we find the giants and the dwarfs, from eagle owls to owlets, eagles to pygmy falcons. There are specialist hunters and generalists, dwellers of savannas, mountains and forest; in fact there are few places in Africa where one will not encounter several of these hunters of the day and the night. The vultures, along with a handful of other raptors, are the avian waste-disposal experts of the African savannas. The fish eagles, osprey and fishing owls specialise in taking piscine prey, and that long-legged terrestrial eagle, the secretarybird, hunts while striding along the ground. Whether it be a pair of Verreaux's eagles taking to the mid-morning thermals, or a common kestrel hovering as it scans the plain, all deserve our admiration.

DIURNAL BIRDS OF PREY
ORDER FALCONIFORMES

Diurnal birds of prey, or raptors, occur throughout the world, the only exception being Antarctica. Of the approximately 300 species a third occur in Africa, many of which are endemic or near-endemic.

Raptors all have sharply hooked, down-curved bills, powerful feet with well-developed claws, and strong wings giving them excellent powers of flight. They range in size from the diminutive 50 g pygmy falcon to the mighty martial eagle with a wingspan of up to 2,6 m. They are active hunters, some being specialists and others more generalists, but several, for example the tawny eagle, are not averse to scavenging and snacking on carrion as well. There are the aerial hunters, the hoverers, the wait-and-pounce specialists, the ground stalkers and the fish snatchers.

The power that some of these birds can exert was brought home to us when we examined prey remains below an active crowned hawk-eagle nest. The skulls of several adult vervet monkeys and blue duikers had neat, round holes through the craniums made by the long, powerful talons of this eagle.

Left: *The common kestrel occurs widely in Africa and Eurasia as far east as the Philippines.*
Top: *Cape griffon.* (Photo: John Carlyon)

Clutch sizes are small, with some species laying just one egg. In the case of several of the eagles, such as Verreaux's, two eggs are laid but the first-hatched and therefore stronger chick kills the other; this is called Cainism. In other species, such as the booted eagle, two young are often successfully raised.

Eagles, hawks and kites

FAMILY ACCIPITRIDAE

Africa is home to 23 eagles representing a diverse mix of sizes, habitat preferences and feeding niches. The most populous genus is *Aquila*, or **typical eagles**, with eight members. In size they range from the martial eagle (wingspan up to 2,6 m) to the diminutive booted eagle (wingspan up to 1,3 m).

Many are African endemics or near endemics, such as Verreaux's and the bateleur which spill over into Arabia, but a number have extensive ranges beyond the borders of the continent. The golden eagle, with a very wide Palaearctic range, has small breeding populations in north-western Africa in Mauritania, Morocco and Algeria. Another species restricted to the north-west is Bonelli's eagle, but it has a much wider Eurasian range. Several eagles are non-breeding migrants from Eurasia. Among them are the most difficult of large raptors to differentiate in the field, particularly in sub-adult plumage: the greater and lesser spotted eagles, the steppe eagle (sometimes included with the resident tawny eagle) and the imperial eagle.

With its great size the martial eagle is able to hunt a wide range of prey species in the savanna woodlands of sub-Saharan Africa, including small antelope such as dik-dik and young steenbok. However, the crowned hawk-eagle with its short but very broad, rounded wings, long tail, incredibly strong toes and long, lethal talons has to take the title of master predator among African eagles. It has been recorded killing prey up to five times its own weight, including young bushbuck of 20 kg, and full-grown forest duiker of several species. Prey that is too large to lift off the forest floor is dismembered and transported in pieces. In East Africa the 5 kg suni is frequently taken, as are tree and rock hyraxes, and monkeys. In a suburban area of South Africa, crowned hawk-eagles nesting in heavily wooded gorges nearby regularly include domestic dogs and cats in their diet, much to the consternation of their owners. Although mammals make up the bulk of their diet, large birds, such as francolins, guineafowl and hadedas, are not infrequently taken. Predation on birds seems to be influenced by the region in which pairs occur. Not only is the crowned hawk-eagle a mighty predator, it is also a nest-builder of great repute. After several years in use, the nest may measure as much as 2 m across and 3 m deep, and may be constructed in a lofty forest giant up to 40 m from the ground.

Verreaux's, or black, eagles require suitable high cliffs on which to construct their massive stick nests. These territorial and resident birds are specialist hunters with their distribution allied very closely to that of their principal prey, the rock hyrax (*Procavia* and *Heterohyrax*). At up to 3,5 kg, these mammals make up more than 90% of the food intake of Verreaux's eagles in most areas. These eagles have also been recorded as taking small antelope such as dik-dik and klipspringer, particularly in

MATOBO RAPTOR BONANZA

There are few places on earth that can come close to the raptor density of the Matobo National Park and surrounding granite hill country in western Zimbabwe.

Among the rounded granite outcrops lie dry semi-deciduous forests and thickets, deciduous woodland communities, open grassland and strips of evergreen forest along the river courses, as well as artificial impoundments. This great diversity of habitats supports an enormous array of prey species and therefore avian predators.

In this rugged area of just 620 km² there are 32 breeding species of raptor, including owls, and 18 non-breeding seasonal visitors. About 76 pairs of raptors breed at any one time during the principal nesting season in each block of 100 km². In one intensively surveyed block no less than 98 pairs were found breeding. This density is unusually high for a population of mainly resident and predatory, as opposed to scavenging, raptors. For large and medium-sized eagles the breeding density in Matobo is probably unique, certainly on the African continent. This is home to 60 breeding pairs of Verreaux's eagles, more than 25 pairs of Wahlberg's eagles and perhaps as many as 45 pairs of African hawk-eagles. Matobo is also the site of the most intensive study ever undertaken on an African raptor, Verreaux's eagle. Because of limited suitable habitat, species such as the crowned hawk-eagle and African fish eagle occur in low numbers.

A survey of diurnal raptors in the 27 km² Lamto Reserve in the Ivory Coast found an incredible average of 47 breeding pairs for each 10 km², yielding an extrapolated average of 470 breeding pairs for every 100 km². Although this density is higher than in the Matobo area, most of these breeding raptors were small or medium-sized species such as the little banded goshawk (*Accipiter badius*) and the African hobby (*Falco cuvierii*). The large palmnut vulture (*Gypohierax angolensis*) and African harrier-hawk (*Polyboroides radiatus*) were apparently feeding mainly on the fruits of oil palms, not on prey.

Less detailed studies of raptor densities in other regions of Africa have come up with very low figures. In a 3 500 km² area of Eastern Province (South Africa) the density of diurnal raptors was found to be as low as four pairs per 100 km². In a similar study in the Embu district of Kenya the researcher found a maximum density of eight breeding pairs for each 100 km². So, without doubt the raptor avifauna of the Matobo comprises the densest and most diverse community of primarily hunting raptors that has to date been described from any comparable area in the world.

Crowned hawk-eagles construct massive nests that are used season after season. (Photo: John Carlyon)

the Horn of Africa. Some pairs take to preying on sheep lambs and goat kids, and although this is rare, it has resulted in their persecution by farmers. They are generally held to eat carrion only rarely, but this is perhaps not quite correct. We know of several cases of these eagles, in particular immature birds, being captured or killed at sheep carcasses that farmers had poisoned or set with gin traps. In the mountains of the Western Cape province of South Africa, this eagle is sometimes called the *tiervoël* (leopard bird), as it will dive repeatedly at leopards and so indicate their presence. This action is applied to caracal and even large domestic dogs.

The short-toed snake eagle, of which two subspecies are African residents, has a wide Eurasian distribution. The brown and banded snake eagles occur only in sub-Saharan Africa. The common name is a clear indication of the most important component of their diet: snakes. Although the banded snake eagle selects small, thin snakes the other species readily take much larger specimens, including puff adders, mambas and cobras up to 3 m long. Smaller snakes are swallowed whole, occasionally while the eagle is in flight, after the head has been

crushed in the bill. Larger snakes are torn apart and eaten in pieces. On occasion snake eagles eat lizards, including large varanids, gamebirds and such mammals as hares. In the central Namib Desert we witnessed a black-breasted snake eagle snatch a Namaqua sandgrouse off the three eggs she was incubating, breaking only one in the split-second action. The Congo serpent eagle, distinctive with long, barred tail and large eyes to aid vision in the dim light of the forest understorey, preys on chameleons, other lizards and probably small mammals in addition to snakes, but virtually nothing else is known about this raptor.

The African fish eagle is probably the continent's most abundant large eagle, with population estimates going as high as 200 000 pairs. This is another raptor that has evolved a specialist diet and hunting technique. It catches, with its feet, fish of up to 3 kg but generally prefers smaller specimens, which are carried on the wing to a favoured tree perch or the shore or bank of the water body. Large specimens are "planed" to the water's edge: something akin to surfing but without the benefit of wave propulsion. Apart from freshwater bodies, the African fish eagle frequently hunts along suitable coastlines. It will also readily hunt birds, such as the fish eagles that prey almost exclusively on flamingos on Kenya's Lake Bogoria. We have observed fish eagles making a number of successful bird kills, but there have been occasions when the intended prey managed to elude the hunter. One fish eagle latched on to an Egyptian goose, but the latter with much calling, wing-flapping and loss of feathers managed to shake off the raptor after dragging it almost 10 m along the ground.

The osprey (*Pandion haliaetus*) is also a predator of fish, equally on fresh and coastal waters. This specialist raptor has a reversible outer toe and the undersides of the toes are covered with spiny protrusions, an adaptation for holding slippery piscine prey. Unlike the African fish eagle which usually takes its prey from close to the surface of the water without getting other than its feet wet, the osprey frequently plunges completely below the surface, even a metre deep.

An unusual member of this family is the African harrier-hawk, or gymnogene, with its long, broad wings, bare facial skin which changes colour according to mood, long legs and a tibio-tarsal joint that can bend backwards; a "double-jointed" bird if you will! This latter is a very useful adaptation for feeling for prey in tree holes, rock crevices and weaver nests. Their prey includes birds' eggs and fledglings, but also on occasion adult birds taken from a nest or nest cavity, as well as lizards of

several species, small mammals including bats, and considerable quantities of insects and other invertebrates. Gymnogenes will spend several minutes exploring crevices on cliff-faces, often maintaining position with slow flapping of the large wings. They also hang upside down from branches holding weaver nests, inserting one foot to check for the presence of eggs or chicks and often destroying the nest in the process. In West Africa oil palm fruits may make up an important component of the diet.

The nine **goshawks** and eight **sparrowhawks** are in many cases difficult to identify, particularly the immature birds. With a few notable exceptions such as the red-breasted, red-thighed and black sparrowhawks, most members of this group can be called "barred hawks", a reference to the markings on their undersides. Many species are difficult to observe, with a fleeting glimpse in woodland or forest all that is often offered, but the three chanting goshawks often perch in exposed positions. In the arid interior of southern Africa the pale chanting goshawk is commonly seen on telephone- and fence-poles.

Many species have a wide distribution, occupying a variety of woodland and savanna habitats, but some, such as the chestnut-flanked and red-thighed sparrowhawks, are small bird-hunting raptors of the tropical forests. Only one species, the northern goshawk, has a very limited range, occurring only in a tiny area of Morocco. The long-tailed hawk (*Urotriorchis*

Above left: *Birds make up almost three quarters of the diet of the African hawk-eagle.* (Photo: John Carlyon)
Above right: *The African fish eagle utters one of the continent's best-known bird calls.*
Left: *Downy young of Wahlberg's eagle.* (Photo: John Carlyon)
Far left: *Black sparrowhawk drying off after bathing.*

Far left: *Adult pale chanting goshawk hunting harvester termites.*
Left: *The European honey buzzard shows considerable colour variations.*
Below: *Young black-shouldered kites on the nest.*

macrourus) of the tropical rainforests has been placed in a separate genus primarily on the basis of its very long tail; otherwise it is similar to the sparrowhawks and goshawks. Many are general hunters of lizards, birds, small reptiles and invertebrates, but the red-thighed, little, Ovambo and red-breasted sparrowhawks concentrate on small birds. Often considered to be the master bird-hunter of this group is the black sparrowhawk, which takes an array of large species such as guineafowl, francolins, turacos, pigeons and hornbills.

Six **harrier** species are recorded in Africa, including two endemic residents: the African marsh and black harriers. There is also a small breeding population of Montagu's harrier and the European marsh harrier in the extreme north-west. Harriers are long-winged, long-tailed and long-legged raptors that occupy open habitats and usually hunt on the wing, flying slowly and close to the ground. In southern Arabia we have watched European marsh harriers hunting fiddler crabs over mangrove mudflats and isabelline wheatears in low scrub nearby. All African harriers are ground-nesters and nearly always roost on the ground.

Buzzards are well represented in Africa, with 12 species. Although most are endemic, the common, or steppe, buzzard is

an abundant non-breeding Palaearctic migrant from northern and eastern Europe. A few enter Africa over the Strait of Gibraltar, but the vast majority stream in over the Suez flyway. Large numbers reach South Africa; in fact it is the most abundant medium-sized raptor in the country during the southern summer. The jackal, augur and Archer's buzzards, sometimes considered subspecies of the same bird, are common residents and frequently seen. In South Africa the jackal buzzard is particularly common in the extensive farming country on the central semi-arid plateau, where they overcome the scarcity of nesting trees by building their nests on windmill platforms. In such situations they are often extremely tolerant of people. A melanistic form of the augur buzzard, which normally has white undersides, is quite common in the higher montane areas of eastern Africa, particularly the Aberdares and Bale Mountains. Most buzzards are opportunistic hunters, taking everything from small mammals, snakes, lizards and birds to a wide range of invertebrates. Several species, such as the jackal and common (steppe) buzzards, readily scavenge from road kills and other sources. A tar road of 80 km from our home village to the nearest small town supports several jackal buzzards and steppe buzzards "in season"; they make a fairly good living from squashed Cape hares, steenbok and other unfortunates. The European honey buzzard could be mistaken for a steppe buzzard at a distance. When migrating into Africa from Eurasia, spectacular numbers crowd the flyways: as many as 120 000 cross the Strait of Gibraltar and possibly 25 000 the Bosporus. However, once in Africa they disperse and only singles or small groups are seen. This rather specialised feeder eats mainly bees and wasps, frequently digging for their larvae in underground nests, and has been recorded taking wasp nests from under the eaves of buildings and from trees with the feet, while in flight.

The **bat hawk**, sometimes referred to as an aberrant kite, occurs throughout tropical Africa. It hunts at dusk for the small insectivorous bats that make up the bulk of its prey. They are taken on the wing and transferred from foot to bill, then swallowed whole. The bats are usually hunted over relatively open areas where they can be seen, but some birds habitually frequent cave entrances with large bat populations. In full darkness the birds roost, although there are records of bat hawks taking advantage of bats attracted to artificial light in towns long after sunset. We know of one record of a bat hawk and a barn owl plucking bats out of the air as they criss-crossed in front of a cave entrance.

The last group in this family are the **kites**, of which five species occur on the African continent. Two, the black-shouldered and swallow-tailed kites, are fairly small species with similar coloration but different tail structures. Black-shouldered kites have a wide African distribution, spilling over into southern Europe and Arabia, and occurring in many different habitats. Out of the breeding season they roost communally in groups of up to 20 birds, and although 500 have been recorded this is exceptional. They apply two methods of hunting the small rodents that make up 90% and more of their prey: dropping from perches or hunting posts, or hovering until a mouse is sighted and then plunging onto it. The perch hunt is the one most frequently adopted and mice of up to 100 g are taken. In contrast the similarly sized swallow-tailed kite, largely restricted to the Sahel savannas, feeds mainly on insects which may be taken on the ground or gracefully on the wing. Black-shouldered kites are solitary nesters but the swallow-tails are colonial nesters, with up to 20 pairs nesting in close proximity.

The *Milvus* kites, of which two species are present in Africa, are large, fork-tailed scavengers. The red kite breeds only in the extreme north-west of the continent. There is some movement southwards out of the breeding season but this is poorly understood and would seem to involve only small numbers of birds. In contrast the black kite, with its subspecies the yellow-billed kite, occurs virtually throughout Africa, where it is a common sight around and in human settlements. It is often considered to be the world's most abundant, adaptable and successful diurnal raptor, having overcome the difficulties attached to living in close association with humans. ■ SEE PAGE 156.

Far left: *The short-toed eagle is considered by some taxonomists to be the same species as the black-breasted snake eagle. This form occurs in north Africa and Arabia.*
Left: *Blond phase of the tawny eagle exhibiting a full crop.*

Right: *Sooty falcons breed in north-eastern Africa and the Middle East, spending the winter as far south as south-eastern South Africa and on Madagascar.* **Below:** *Amur falcons often roost in large flocks in tall trees in their wintering grounds.*
(Photos: John Carlyon)

Above: *Adult lanner falcon in the Drakensberg of South Africa.*
(Photo: John Carlyon)
Left: *Red-necked falcons occur widely in sub-Saharan Africa as well as in India.*

Falcons and kestrels

FAMILY FALCONIDAE

Of the 21 African members of this diurnal raptor family, all but two fall within the genus *Falco*. They range in size from the very small pygmy falcon (average 60 g) to medium-sized birds. All are characterised by strongly hooked and notched bills, proportionally long and pointed-tipped wings and long, slender toes with strong, sharp talons. In most cases the females are larger than males. Some species show considerable sexual dimorphism, whereas others do not. Unlike the Accipitridae, the Falconidae do not eject the faeces in a powerful liquid stream (or "slice") but drop it straight down from the cloaca (known as a "mute"). None of the birds in this family build their own nests but may usurp those of others, or lay eggs directly on the ground or a rock ledge. In this characteristic, and the fact that vertebrate prey is killed by biting and breaking the neck, they bear some similarity to owls.

Many kestrels hunt by first hovering at no great height, seeking out prey, and then diving onto it and grasping it in the foot or feet. Most falcons usually take birds (occasionally bats) on the wing, either in pursuit or after a steep dive (a "stoop") from height with the wings drawn into the body. The peregrine falcon is the classic stoop-hunter, with speeds in almost vertical dives of 250 km/h being attained, but it also catches birds after a horizontal chase, and on occasion from below the target.

Although some species only occur in Africa (the pygmy and Taita falcons, the African hobby, and the fox, greater and Dickinson's kestrels), others have both resident and non-breeding migratory populations (such as the peregrine falcon), and five others are non-breeding migrants from the Palaearctic Region. It is in fact some of the migrants that extend over the greatest area of the continent.

Some species, such as the common kestrel and red-necked, lanner and peregrine falcons have extensive African distributions, but a few are very localised. The Taita falcon, resembling a diminutive peregrine, has apparently separate populations, one centred on Zimbabwe and the other on Kenya. Then there is Eleonora's falcon, which breeds almost exclusively on coastal islands; non-breeding migrants extend in a narrow belt along the Mediterranean, Red Sea and Indian Ocean coastlines. Dickinson's kestrel occurs in a broad belt to the south of the equator, particularly where wild palm groves are present.

The lesser kestrel breeds very widely across Eurasia, from southern Europe to China and with a small population in extreme north-western Africa. Migrants enter Africa across a broad front, crossing the Sahara and dispersing over the savan-

---

---

nas, with the greatest overwintering populations concentrated in South and East Africa. They are nearly always present in flocks and night-time roosts may accommodate hundreds of birds. Distances covered on these migrations are great: more than 8 000 km in some cases. Red-footed falcons also enter Africa on a very broad front across North Africa and the Sahara, but most cross between the Italian peninsula and the Middle East. They pass through much of Africa to reach the main overwintering range in southern and south-central Africa. Roosts, often in large trees in towns or villages, may be made up of several thousand falcons. ■ SEE PAGE 156.

Secretarybird

FAMILY SAGITTARIIDAE

The name "secretarybird", generally believed to be derived from the cluster of quills on the head, is in fact a corruption of the Arabic *saqr-et-tair* which means hunting-bird.

Sometimes referred to as a terrestrial eagle, the secretarybird spends more time hunting on the ground than any other African raptor, but it is capable of strong flight. This large, eagle-like bird may reach a total length of 1,5 m and stand approximately 1,3 m tall. It has extremely long legs but rather weakly developed feet which lack the grasping ability of many other raptors. This sub-Saharan endemic is only absent from forested areas but shows a marked preference for open savan-

na, plains and tracts of semi-desert. Although much of the day is spent foraging on the ground, it will seek out shade when midday temperatures are high, and at night roosts in trees, frequently in the nesting tree. Although pairs generally forage separately, they not infrequently roost together at night. Up to 50 secretarybirds have been recorded close to each other in the vicinity of water in arid areas, but such gatherings are rare.

All hunting is done on the ground and secretarybirds take a very wide variety of prey, ranging from insects to lizards, snakes, young tortoises, nestlings and small rodents. The feet are brought into play when a prey animal is located, to hit down on it swiftly and hard. In Lake Nakuru National Park in Kenya we watched a secretarybird pound a small dry bush into fragments, then retrieve from it a dead *Otomys*, a fairly large, short-tailed grass rat. It proceeded to swallow it head first. On several occasions we have watched these birds near our Karoo home catching various species of lizard, including agamas, skinks and sand lizards. They either catch them with little chasing or they zigzag at speed with wings half open, but in these cases their prey capture success seems to be rather limited. They are reported to kill large snakes, such as puff adders, when encountered but the quantity of snakes they catch is certainly exaggerated. It has been reported that they may take the very young of steenbok and other small antelope but we have not been able to pinpoint definite proof of this. They certainly eat large quantities of locusts and will remain close to swarms until they are satiated.

Left: *Secretarybird in the process of swallowing a mouse.*
Below: *Cape griffons are found only in southern Africa; populations have declined drastically during this century.*
(Photo: John Carlyon)

Above: *Two separate populations of the pygmy falcon occur, one in arid south-western Africa and the other in East Africa.*
Right: *The Mauritius kestrel, restricted to the island of the same name, hovers on the brink of extinction, as do many species on islands around Africa.*

Far left: *The slender bill and predominantly white plumage distinguish the Egyptian vulture from all others.*
Left: *The unmistakable secretary-bird has been described as a long-legged terrestrial eagle.*

The nest is an enormous structure of sticks that may measure as much as 2,5 m across and 0,5 m thick. It is constructed on top of a bush or low tree and is frequently used for several years in succession. Over time these structures become so heavy that they break and bend branches, causing the nest to settle well below the tree's uppermost branches. The female lays between one and three eggs and incubation takes up to 45 days. Both parents feed the young with regurgitated food, prey not being carried to the nest. Larger regurgitated items such as rats and snakes are first torn up by the parent at the nest and fed piece by piece to the young. As the youngsters get stronger, food is often regurgitated at the nest and left for them to feed alone. Once they have left the nest the juveniles accompany the parent birds. ■ SEE PAGE 156.

Vultures

FAMILY ACCIPITRIDAE

There are 11 species of vultures and griffons recorded in Africa, many of which typify the "classic" vulture: nearly bare head and neck, large size and feeding on carrion. However, several do not fit so neatly into this classic form. There is the palm-nut vulture, also called the vulturine fish eagle, which is considered by some to be the evolutionary link between the fish eagles and the Egyptian vulture. It is a hunter of fish but mainly consumes a very un-raptor-like diet of oil-palm and other fruits. However, its skeletal structure most closely resembles that of vultures. One of our favourite spots on the African coast lies close to the borders of Kenya and Tanzania, and it is here on long coral reefs with baobabs dominating the vegetation that we have watched several of these vultures walking over the mudflats at low tide catching crabs. After local fishermen had cleaned their catches of the day, these attractive birds move in and glean scraps from among the canoes. They have also been recorded feeding on a variety of birds and small mammals. In the botanical gardens of Entebbe we saw one of these raptors sharing roof space with a hamerkop, both busy preening and showing not the least interest in the other. Palm-nut vultures are resident and pairs defend a limited area around the nest site. Only a single egg is laid and incubation takes about 44 days.

Another rather un-vulture-like vulture is the bearded vulture, or as it is sometimes called the lammergeier (lamb vulture) – an erroneous reference to its supposed depredations. This massive bird, with a wingspan that can exceed 2,8 m and a long wedge-shaped tail, takes its name from the distinctive dark, bristly "beard" that extends down from each side of the bill base. In Africa it occurs mainly in the mountainous areas associated with the Great Rift Valley and the high Atlas range of the north-west. It is now rare in the south but still relatively common in the Ethiopian Highlands. Throughout its limited African range it seldom descends lower than 2 000 m above sea-level but in the harsh Danakil Desert it has been recorded under 300 m. Elsewhere it occurs widely in Eurasia but particularly in the west its numbers have been greatly depleted. Although through much of its range single birds and pairs are usual, in Ethiopia more than 20 individuals may be seen together. In that country it commonly frequents human settlements in search of edible scraps, showing little fear of people. It is probable that the population in southern Africa is isolated from more northerly populations, but given their strong powers of flight some movement cannot be ruled out, particularly as there are occasional sightings in the highlands that straddle the border between Zimbabwe and Mozambique. Any dispersal almost certainly involves immature birds as the adults seem to be restricted to a fixed home range.

The bearded vulture feeds on carrion, often in the company of other vultures, buzzards and ravens, but unlike other species it also feeds on bones, fragmenting them by dropping them on rock anvil sites, usually from heights of up to 150 m. These sites, known as ossuaries, may be in use for many years, even centuries, and the rocks become carpeted with bone fragments. Bone dropping is executed with great accuracy and the descent to retrieve smashed fragments is rapid, particularly when ravens and kites are in the vicinity, ever ready to snatch a free meal. One of the most remarkable aspects of this unusual diet is that bones of up to 25 cm long may be swallowed whole! There are also records of bearded vultures lifting and dropping tortoises on the "anvils". When they locate a carcass they may carry food to a ledge or nest site in the crop, regurgitate it and return for more.

The nest, often used by the same pair for many breeding seasons, is usually located in a cave or deep overhang. Depending on the location, nests may be from 3,5 km to 30 km apart, which indicates how much the size of home ranges differs. The huge nests, up to 3 m across and 1 m deep, are built mainly of sticks but are liberally "decorated" with bones, dried skin, other vegetation and even clothing. Although two eggs are usually laid, only one chick is raised. Apparently Cainism is not involved; presumably the last chick to hatch is unable to compete successfully with its older sibling for food.

Another strange, untypical vulture is the Egyptian, otherwise known as Pharaoh's chicken! This bird, yet again a taxonomic paradox, may be related to the palm-nut vulture; in behaviour it more closely resembles the bearded vulture, but it does not look like either of them! This is a rarity in that it is a tool-using bird, using stones held in the bill to crack ostrich eggs to gain access to the nourishing content. Eggs of such species as pelicans and flamingos are grasped in the rather slender bill and hurled on the ground until they break. It also scavenges on scraps around carcasses, will feed on termites and other insects but has the rather indelicate habit of dining on human faeces. Indelicate as this may be, it serves a useful purpose in cleaning up material that could otherwise be a reservoir of disease in a continent where toilets are in extremely short supply and the back of a bush or rock must serve.

In the Dhofar of Oman Egyptian vultures frequent rubbish dumps and we have watched them brazenly snatching scraps from around fishing nets and in coastal markets, not infrequently competing with sooty gulls and non-breeding gull species. This vulture is now extremely rare in southern Europe, still occurs in parts of Asia, and in Africa is widespread but patchily distributed.

Another slender-billed species is the hooded vulture, an African endemic. It is generally solitary except at carcasses, where it is unable to compete with the large vultures, so it circulates and picks at scraps on the periphery. In West and northeastern Africa this vulture is a common scavenger around human settlements. In Entebbe we watched several of these birds lined up along the roof of one of the town's top hotels, all peering down intently at the handful of tourists basking at the edge of the swimming pool. What is interesting about this vulture is that it is the only member of this species group than can live in areas of very high rainfall, where it is strongly tied to the wastes produced by human settlements. Although this is not a

Left: *The bearded vulture occurs in mountainous country along the line of the Great Rift.* (Photo: John Carlyon)
Above: *Adult bearded vultures are the aristocrats of the high peaks.*
Opposite page left: *Lappet-faced vultures and a single white-headed vulture (left).* (Photo: John Carlyon)
Opposite page right: *Cape griffons are cliff-breeding vultures, some such colonies being hundreds of pairs strong.* (Photo: John Carlyon)

DECLINE OF THE CAPE GRIFFON

The Cape griffon (*Gyps coprotheres*), the most intensively studied African vulture, has seen its overall population decline dramatically during the course of this century. An estimated 10 000 of these griffons survive, of which some 4 000 constitute breeding pairs. Once a common sight over its southern African range, well over 100 000 of these majestic birds once soared on the thermals over semi-desert and savannas.

Its present classification as rare is on the basis of the precipitous population decline and the continuing threats to its long-term survival. In 1982 it was estimated that between 1 900 and 2 500 breeding pairs survived, but over the intervening 16 years some small breeding colonies have been abandoned and numbers at several large colonies have declined. Perhaps as much as 75% of the surviving population breed in the northern provinces of South Africa, in what

used to comprise the Transvaal. Fewer than 30 breeding sites are known to be in regular use today.

The dramatic decline in griffon numbers has been ascribed to a number of factors. In 1896 huge numbers of domestic cattle died during the great rinderpest outbreak, and this is believed to have resulted in a severe reduction of food available to these scavengers in following years. As a result griffon numbers declined. Of course this was preceded by the disappearance of the great herds of game that once traversed the interior plains of southern Africa. Gradually griffon numbers picked up but by the 1960s colonies had again been abandoned and numbers declined.

In recent decades improvements in animal husbandry and limitation of stock losses meant that the availability of carcasses suitable for scavenging was greatly reduced. Of

course, direct persecution was and still is a major threat, despite the griffon's protected status. Poisoning of carcasses, shooting and disturbance at breeding and roosting cliffs have had serious consequences for this magnificent bird. Chick mortality has also been high because of the shortage of bone fragments to supply adequate quantities of calcium for skeletal development. In the past such fragments would have been provided by the bone chewing of such predators as lions and spotted hyaenas, but except in the few major game parks these carnivores are but a memory.

In order to try to slow mortality and declines there has been a great deal of publicity for the Cape griffon, as well as the establishment of "vulture restaurants" where they feed and obtain bone fragments for the chicks, and measures to reduce mortality resulting from collisions with power-lines.

colonially breeding species, its tree nests may be built in loose clusters over a fairly limited area.

Now we come to the "true" vultures, those that we can all instantly recognise as the consorts of witches and wizards, with near-naked heads and necks dripping blood and gore ... Actually they serve as the dustmen, or more politically correct waste disposal experts, of the African savannas and semi-deserts. They are all large, with heavy, powerful bills designed for tearing flesh and gut, and mighty wings for soaring high and long on thermals in their tireless quest for carcasses on which to feed.

There are those such as the lappet-faced and white-backed that are tree-nesters, often solitarily but on occasion in loose colonies because of limited numbers of suitable nesting trees. Then there are the colonial cliff-roosters and nesters, the Cape and Rüppell's griffons, signposted by their faecal whitewashing

of the sheer rock walls. Both of these cliff dwellers are African endemics. The Cape griffon of southern Africa is now greatly reduced in numbers but Rüppell's is still abundant, with its populations probably much as they have been for centuries. Rüppell's griffon, with its "scaled" plumage and ivory-coloured bill, is distinctive and cannot be mistaken for any other species. This is the dominant vulture of the Sahel and East Africa wherever there are suitable roosting and breeding cliffs. Foraging up to 100 km away on the surrounding savannas and open woodlands, up to 1 000 pairs may breed on suitable cliffs and sheer mountain faces. One of the best known and most accessible breeding sites is in Hell's Gate National Park above Lake Naivasha in Kenya.

Another cliff-breeder but in much smaller colonies is the Eurasian griffon, a fairly common vulture in some North African mountainous regions. Although still fairly common in its North African range, western Asia and Arabia, in Europe it has been greatly depleted. Adding to resident populations, European birds migrate into north-western Africa and larger numbers enter Africa over Suez.

Largest of the resident "true" vultures, although with populations in Arabia, is the lappet-faced with a wingspan that can exceed 2,6 m. The somewhat similar cinereous vulture, which however lacks the pinkish facial skin, is a rare non-breeding winter visitor to North Africa. The lappet-faced dominates other vultures at carcasses and unlike the others is able to break into the body cavities of dead animals relatively easily. It reaches its highest numbers in semi-desert areas but even then the maximum group size is no more than 50, and only for short periods, usually at a food source or water. When feeding, this vulture will hold a carcass with its feet and tear the flesh and skin with its large and powerful bill. Unusual for a vulture, it has been regularly observed actively attacking, killing and eating adult flamingos, chicks and eggs in breeding colonies. It also readily feeds on emerging termite alates as well as locusts. The stick nest is huge (up to 3 m across) and so heavy that it can cause the structure to collapse or bend the tree's supporting branches. After leaving the nest the young depend on the parents for food for up to two months. Young birds, although

unable to compete with adults of their own kind at carcasses, are able to dominate griffons.

One of the most frequently seen and most abundant African vulture in the savannas is the white-backed. At carcasses they may gather in their hundreds, squabbling, screeching and lunging at each other with wings outspread. When gorged they withdraw from the carcass, frequently sitting on the ground with bulging crops or otherwise flying into nearby trees, particularly if large mammalian predators such as lions are in the close vicinity. These and other vultures frequently bathe, particularly after feeding, really getting themselves well soaked and then standing with open wings to dry in the sun. A pair may nest well separated from others but in favourable areas several nests may be located in loose association. In this situation white-backed vultures show little territoriality, only defending a small area immediately around the nest.

White-headed vultures tend to occur at low densities throughout their range. They frequently arrive at carcasses before other vultures and griffons but being unable to compete with them when the squabbling begins they pick up scraps, pieces of skin and sinew. Unlike most other vultures they actively hunt, taking flamingos, chicks and eggs at breeding colonies. ■ SEE PAGE 156.

Far left: *White-backed vultures are abundant and widespread.*
Above: *Marsh owl wing detail.*
Left: *Although the marsh owl is usually solitary or lives in pairs, groups of up to 40 have been recorded.*

Far left: *Wood owls have a wide sub-Saharan range but are easily overlooked.*
Above: *The small African barred owlet hunts mainly at night but will also seek its quarry during daylight.* (Photo: John Carlyon)
Left: *Barn owls produce more eggs and raise more young during periods of food abundance.*

NOCTURNAL BIRDS OF PREY: OWLS
ORDER STRIGIFORMES

When the diurnal birds of prey go to roost, the hunters of the night, the owls, emerge, sweeping on silent wings over savanna, desert, forest, woodland and mountains. There are few places on the African continent where owls do not hunt, breed and add a hint of mystery or, to some people, foreboding. They range in size from the tiny pearl-spotted owl to the giant Verreaux's eagle owl. All have soft plumage, large forward-facing eyes, a hooked bill and powerful feet, with each toe equipped with a powerful, sharp talon. The ear openings, not apparent because of a covering of feathers, are very large when compared to those of many other birds. Unlike the diurnal raptors they have no crop in which to hold food, but like all raptors they regurgitate pellets of undigested material such as hair, feathers, bones and insect exoskeletons. Birds, unlike mammals, do not chew their food but swallow it whole, or tear it into pieces small enough to swallow. The narrow pyloric opening of birds allows the passage of only small food items, and

the absence of free acidity in the stomach prevents complete digestion. Stomach acidity is particularly weak in owls. Hard parts of the food items are usually enclosed in softer material such as hair and feathers. This results in little damage to prey bones, and the content of owl pellets are therefore useful indicators of their diet.

The combination of superb vision and hearing plus silent flight makes owls the master hunters of the night. Depending on the species they hunt a very wide range of prey, which includes invertebrates, rodents, birds, bats, geckos and mammals such as hares, hedgehogs and even small carnivores. Prey is grasped in the feet, usually on the wing. Although owls, because of their nocturnal habits, are not frequently seen unless searched for at their daytime roosts, each has a characteristic call that serves as its identity tag, from the deep hoots of some of the eagle owls, to the chirrups of the scops owls and the eerie, somewhat maniacal screeching of the barn owl.

Of the two owl families in the world, the Tytonidae (barn owls) have two species in Africa. The Strigidae ("typical owls") include 40 different species in Africa, although some authorities recognise fewer.

Because they fly mainly at night and often live in places difficult of access, many of Africa's owls are poorly known. Some have been seen on only a few occasions. One classic example from the forests of eastern Democratic Republic of Congo and adjacent Burundi is the Congo bay owl, which is known from no more than three instances: a single specimen, one captured and released bird, and one possible sighting. Some consider it to be conspecific with the oriental bay owl (*Phodilus badius*) but the physical distance separating their two ranges makes taxonomists nervous about putting them together under the same umbrella. Also, there are several differences between the two that make some taxonomists comfortable about keeping them separated.

Other African owls about which little information exists include the sandy scops owl, known from very few localities across the length of the tropical lowland rainforests. Other tropical forest dwellers are Fraser's eagle owl, Shelley's eagle owl, the Akun eagle owl, rufous fishing owl and Albertine owlet, which is known from a mere handful of specimens. The Sokoke

scops owl is known only from the forest block of the same name in coastal Kenya, where there are an estimated 1 500 resident pairs. Another denizen of tropical lowland forest is the mysterious maned owl, known from a few isolated records from Liberia to western Democratic Republic of Congo. Then there is the chestnut-backed owlet and the outlier population of Hume's tawny owl, known from a limited area in the Egyptian Red Sea Hills. Other than the diet of a number of species, we know very little about Africa's owls outside southern Africa.

Barn owls

FAMILY TYTONIDAE

This family encompasses only two species on the African continent, the barn owl and African grass owl. The barn owl, one of the most widespread of the Strigiformes, is commonly associated with people and their buildings. In Africa they frequently rest and breed in rock crevices, hamerkop nests and tree holes (particularly favoured are baobabs), but also in buildings. Like many owl species, they use the same roosts and breeding sites year after year. Several such sites with which we are familiar have massive accumulations of bones and skulls

Right: *The barn owl has an almost worldwide distribution.* (Photo: John Carlyon)
Below: *Little owls occur widely in northern and north-eastern Africa.*
Opposite page left: *The white-faced scops owl eats mainly rodents up to the size of a bush squirrel.* (Photo: John Carlyon)
Opposite page right: *The common scops owl is about the size of a palm dove.*

of small rodents, insectivores and reptiles. Each rainy season the pellets are broken down, leaving only the bones. Below roosts in rock shelters or overhangs these bones accumulate in layers, covered by wind-blown sand and dust, protected from the elements. These deposits, often hundreds and even thousands of years old, are of great value to palaeontologists and archaeologists. The presence of, for argument's sake, the remains of a moisture-loving mouse in a desert environment indicates that a climatic change took place in the area.

Although small rodents and insectivores make up the bulk of the barn owl's diet, it also hunts bats, often at cave entrances where large numbers roost. Birds are also hunted. One pair of barn owls in coastal Namibia took large numbers of small wading birds from the mud- and sand-flats. Many of their pellets that we collected were the normal size and shape but the legs of the prey birds were intact and stuck out well beyond the pellet. A study in Mali, West Africa, found several pairs of barn owl to be feeding almost exclusively on frogs and toads.

The very similar but larger and darker grass owl has a rather patchy distribution, mainly occurring south of the equator. The principal limiting factor is the need for moist grass-lands which provide adequate cover and prey. Small rodents are by far the most important food item but prey up to the size of hedgehogs and young hares has been recorded. ■ SEE PAGE 159.

Typical owls

FAMILY STRIGIDAE

Many but not all members of this family have tufts of feathers on either side of the head, frequently called "ear" tufts. Of the tiny scops owls only the common and the white-faced have wide distributional ranges, with the remainder occupying very limited areas. All but the white-faced scops are less than 100 g. In the case of the common scops owl, some separate it into two species, a sub-Saharan and a North African, with the latter extending into Eurasia. The former is resident but the latter has both a resident and migratory population. Apart from minor differences in the physical appearance of all of the scops owls, the best method of identifying them is by each's distinctive call. The chirrup-trill of the southern form of the common scops owls is one of our "most favoured" African

calls. The plumage of scops owls offers them superb camouflage against the bark of trees, where they spend the daylight hours. Insects and other invertebrates make up much of their prey but some species will also take small rodents, shrews and passerine birds.

Another group of very small owls, belonging to the genus *Glaucidium*, are represented by at least five endemic species in Africa. They are stocky, round-headed owls that are often seen hunting during the day, particularly at dusk and dawn. The pearl-spotted owl, sometimes referred to as an owlet because of its size, is the most diurnal of all. We have seen them hunting butterflies in mid-morning, plucking them from the air. Insects and other invertebrates are the most important component of its diet but lizards, frogs, birds and small mammals are also hunted. All nest in tree holes, even those excavated by woodpeckers and barbets. Another small owl, the appropriately named little owl, occurs in a very wide Eurasian range, as well as throughout Africa north of the Sahara and even on some of the massifs in the centre of that great desert. They also occur along the Red Sea Hills and into northern Somalia and the Ogaden of Ethiopia, occupying a wide range of habitats, including desert, particularly where there are rock cavities and crevices to provide shelter from the heat. In one area of flat,

open gravel plains we observed a pair that occupied a hole in a heap of gravel and sand pushed by a road repair crew. Although they eat mostly insects, they gladly take small rodents, lizards and young birds if available. Most hunting takes place at night but they commonly stand in the entrances to their shelters during the day, and will fly readily if disturbed.

The eagle owls are the supreme nocturnal hunters. All are large and have well-developed "ear" tufts. The recognised number of species occurring in Africa ranges from seven to nine depending on one's taxonomic sympathies. Most are African endemics, with the exception of the Eurasian and spotted eagle owls. They occupy many different habitats including montane grassland, savanna (even in arid areas) and forest. Several are poorly known, particularly those associated with forest. Those of more open country, such as Verreaux's and the spotted eagle owl, are better understood. Verreaux's, sometimes called the giant or milky eagle owl, has a length well over 50 cm and rather alluring pink eyelids. It has a broad diet, and its large size enables it to take mammals the size of hares, spring hares, cane rats, hyraxes, galagos and young monkeys. It is also one of the principal predators of hedgehogs, and fairly substantial accumulations of the skins of these prickly insectivores gather below favoured perches. It also hunts many differ-

Opposite page: *The spotted eagle owl is one of Africa's most frequently seen owls.*
Far left: *Mackinder's, or the Cape, eagle owl has a disjunct distribution from south-western to eastern Africa.*
Left: *The "ears" of the long-eared eagle owl are just feather tufts.*
Below: *The Eurasian eagle owl occurring in North Africa tends to be paler and more reddish than more northerly forms, and it is sometimes regarded as a distinct species.*

ent bird species, including other owls, herons, francolins, bustards and a wide array of smaller birds.

The spotted eagle owl occurs virtually throughout sub-Saharan Africa and into southern Arabia, except in tropical forests. This is the "eared" owl sometimes seen sitting on telephone- and fence-posts, and even on the road surface, at night. Unfortunately it is most commonly seen dead on the road, having been blinded by oncoming headlights and struck by a vehicle. However, the large numbers of fatalities in some areas give an indication of how common this owl is. Several pairs occupy the commonage in and around our home village on South Africa's central plateau, each pair keeping within its own territory. At their daytime roosts they are regularly mobbed by other birds; in their turn they harass feral cats and small-spotted genets at dusk. They have an extremely varied diet and will tackle almost anything they can overpower.

Mackinder's eagle owl, known in the south as the Cape eagle owl, occurs in a long but rather narrow swathe of mountainous country from south-western South Africa to 4 300 m above sea level in the Ethiopian Highlands. Much of its range lies along the Great Rift Valley. In the south it is similar in size to the spotted eagle owl, but northern birds become progressively larger; the race in the Ethiopian Highlands is almost as large

as the Eurasian eagle owl, arguably the largest owl in Africa. Although Mackinder's eagle owl does take a few birds, more than 90% of its diet is composed of mammals, up to the size of hyraxes, hares, red rock rabbits and civets. There are also verified records of this owl killing the young of klipspringer and common duiker. However, like most predators they will kill and eat whatever is available, and so crabs make up a substantial percentage of their food in some areas.

Eagle owls are primarily ground-nesters, in a bare scrape among boulders or vegetation with accumulations of pellets and prey as the only "decorations". One exception is Verreaux's eagle owl, usually laying its eggs in the abandoned nests of other birds such as large diurnal raptors or hamerkops.

The most specialised by far are the three species of fishing owl (genus *Scotopelia*). All are endemic to Africa. Unlike other owls, they can be heard quite clearly when flying. The large feet are armed with substantial claws and the underside of the toes have numerous sharp-pointed scales for gripping their slippery prey. They are restricted to riparian woodland in the tropics. The rufous and vermiculated fishing owls are very poorly known; only the more widely distributed Pel's fishing owl has been studied. Its far-carrying hooting is one of the most evocative calls of African waterways. ■ SEE PAGE 159.

6

Sandgrouse, doves and fruit-eaters

This chapter covers a rather disparate group of largely unrelated species, although the sandgrouse and doves may be distantly related. Sandgrouse are birds of savanna and arid lands, but the majority of species covered here are found in woodland or forest. All provide visual pleasure: hundreds of sandgrouse flying in to water with almost clockwork precision, African grey parrots squawking their way to mighty communal roosts, or the sun-bathing antics of mousebirds. Parrots and their kin have played a long and intimate role in a number of human cultures, unfortunately to the detriment of a considerable range of species. Some pigeons and doves are persecuted because of depredations on certain subsistence and commercial crops, and mousebirds are vigorously "controlled" because of their taste for orchard blossoms and fruits. In our eyes, the turacos and go-away birds epitomise the African tropics. Having had the good fortune to have observed several species of turaco, both in captivity and in their natural habitats, we have to admit that they are among our favourite birds. Although the forest species tend to be difficult to observe, their characteristic calls draw the attention of the human wanderer. But, be it humble dove, raucous parrot, agile mousebird, desert-adapted sandgrouse or colourful turaco, all add something to the avian splendour of Africa.

Doves and pigeons

FAMILY COLUMBIDAE

This family includes some of the world's most familiar birds. Who has not seen the street-wise feral pigeon, or doves in parks and gardens? Almost 300 species are recognised worldwide, of which 41, many of them endemics, range through Africa. Some species occur over vast tracts of the continent, such as the Namaqua and palm doves; others are restricted to limited ranges. All have plumpish bodies, proportionally small heads and short, characteristic bills that are often slightly bul-

bous at the tip and have a fleshy cere at the base where the nostrils are located. The feathers are soft and dense, and plumage coloration is virtually the same in both sexes. Some species are seen mainly singly or in small groups, whereas others may congregate in large flocks, particularly outside the breeding season. In agricultural areas some species are considered a nuisance, particularly on grain crops. On occasion insects may be taken by a few of the doves but seeds, fruits and other plant parts are their usual diet. They have large crops and powerful gizzards that act as highly efficient "grindstones".

Left: *Female black-faced sandgrouse; this is an East African species.*
Above: *The double-banded sandgrouse comes to water after sunset.*

The question is often asked, what in fact is the difference between pigeons and doves? Actually nothing of scientific importance. It is an almost arbitrary division, with the term "pigeon" being applied to the larger species, and "dove" to the smaller species. Having said that, we nevertheless speak of rock and stock "doves", rather than pigeons, although these birds are large.

Pigeons and doves build what one can call a minimalist nest, usually a sparse, untidy platform of sticks and grass, on which the one or two white eggs are laid. Looking at the nest from below it is often easy to make out the eggs through the meagre bed. Some, such as the speckled pigeon, construct their nests on rock ledges and not infrequently on or in buildings. The young, known as squabs, are near naked and with the eyes closed on hatching. The first food that the hatchlings receive is a product of the parents' crop and is known, appropriately, as "pigeon's milk". This has been analysed and has been found to be similar in food value to mammals' milk. Both parents take responsibility for nest construction, incubation and feeding the squabs. Although the clutch is small, this is often compensated for by two or more clutches being laid in a season.

The rock dove is the progenitor of domesticated and feral pigeons, and its association with humans goes back many centuries. It is known that the ancient Romans were using these birds to relay messages at least 2 000 years ago, and today pigeon fanciers breed birds for racing, showing and eating. Pigeon breast, or squab, baked in pastry has to be one of the greatest culinary delights! Various species of dove are kept for decorative purposes, and many of the colourful tropical species are popular in bird collections.

One very distinctive group of four pigeons occurring in Africa are the exceedingly handsome green-pigeons, arboreal birds that do most of their fruit-seeking in trees and bushes and seldom venture to the ground. Although the African and Bruce's green pigeons are recognised by taxonomists as distinct species, the validity of the Pemba and São Tomé birds is in some doubt. Many subspecies of the African green pigeon have been described, based mainly on small size and colour differences. This green-pigeon occurs in forests and heavily wooded savannas in the tropics, sometimes penetrating unsuitable habitats along linear gallery forest. Bruce's green-pigeon occupies the drier woodland savannas across West Africa, eastwards to Somalia and Ethiopia, and extending into south-western Arabia. Where the ranges of the two species overlap they may temporarily form mixed flocks, particularly at fruiting trees. When these birds are feeding they climb around among the branches in the manner of parrots. They have complex, soft calls that are very un-pigeon-like and which we find very attractive. On a recent trip to the coastal forests of Omani Dhofar, where Bruce's green-pigeon is fairly common, we watched large numbers of these birds as they feasted on wild figs.

A large number of Africa's doves and pigeons fall into the subfamily Columbinae, and they are all ground-feeders that take mainly seeds. Three groupings can be recognised: the so-called quail-doves and bronze-wings; the rock doves and wood-pigeons (principally the genus *Columba*); and the more advanced turtle-doves (the genus *Streptopelia*). Nearly all have variations on the cooing theme, and their courtship includes a display flight in which the wing-tips are clapped together to produce a surprisingly loud sound.

Above left: *Squabs of doves and pigeons, in this case the speckled pigeon, would not win any beauty contests.*
Above right: *Mourning doves rarely form flocks except at a plentiful food source.*
Left: *The tambourine dove is either solitary or occurs in pairs, occupying forest and dense woodland.*
Far left: *Speckled pigeons occur widely in sub-Saharan Africa.*

The African green pigeon is one of four species occuring in Africa and on adjacent islands.

Emerald-spotted wood-doves are common and are best known for their distinctive call.

The five species of *Turtur* doves are all African endemics which are mainly distributed through the tropics. All are small and are distinguished by the metallic green, blue or violet spots on the wings. Probably the best known in East and southern Africa is the emerald-spotted wood-dove, with its green wing spots and distinctive call that has been likened to "my mother is dead, my father is dead, and all my children are dead, du, du, du ...". Not a very pleasant interpretation for such a lovely bird. Three of the species are forest dwellers, and two are found in wooded savannas.

Unlike most members of this family, male and female Namaqua doves differ in coloration and markings but both have the long pointed tail. This is Africa's smallest dove but has one of the largest ranges, which is continuing to expand both on the continent and through the Arabian Peninsula. Foraging is usually solitary or in pairs but large flocks may gather at waterpoints or rich feeding grounds. This dove is subject to much local and long-distance movements in response to dry and wet seasons.

Pigeons of the genus *Columba* are quite large, and their bare facial areas are often brightly coloured. Most are forest dwellers, such as the three bronze-naped pigeons, the olive-pigeons and the afep pigeon. Then there are those that occupy mountainous, hill and rocky country, with the widespread speckled pigeon being the best known. In West Africa it occupies areas dominated by palms and baobab trees.

A few mainland pigeons occupy small distributional areas. For example, the Cameroon olive-pigeon is restricted to dense, high-rainfall forest on and immediately around Mount Cameroon, and in apparently low numbers. White-naped pigeons are known from the belt of mountains in eastern Democratic Republic of Congo, which includes the Ruwenzoris, and strangely from western Cameroon, where it is

known from a single locality. The very large common wood-pigeon, with a wide Eurasian range, has a substantial resident population in far northern Africa, and this is supplemented by thousands of birds entering from Europe. The stock dove has a more restricted Atlas distribution; it is present in small numbers but there is some seasonal influx of European birds. One of Africa's loftiest dwellers is the white-collared pigeon, endemic to the highlands of Ethiopia from 1 800 m to 4 400 m. It commonly associates with people and their settlements, showing little fear – a strong feature of a number of bird species in that country. The rock dove, one of the true survivors and "exploiters", occurs patchily across northern Africa, even in the Saharan mountain ranges. They are said to gather in vast flocks, up to 20 000 strong, at watering places in Mali.

Most members of the genus *Streptopelia* are smaller than the *Columba* pigeons, and they include the most abundant and best known doves. They usually have longish, white-tipped tails, overall coloration is grey or greyish-brown and most have a partial black neck collar. All are ground-feeders but they perch and roost in trees or bushes. Several species have extensive sub-Saharan distributions but some occupy very limited areas. Eurasian collared doves extend over much of Europe and Asia, and range expansion has been massive; however, so far they have only penetrated Africa in the vicinity of Cairo and the Nile Delta. Sightings elsewhere in the north of the continent and their great success indicate that most suitable habitats will be occupied in coming years. One of Africa's most successful doves is the palm, or laughing, dove. This "international" bird is also spreading its influence and making its presence felt, both naturally and as a result of introduction. Unlike many other *Streptopelia* doves it normally forages on its own or in small parties, with large flocks only congregating at water-points or occasionally at roosts. ■ SEE PAGE 158.

Above: *Mousebirds, here a speckled, are unique to Africa.*
Right: *The red-faced mousebird is distinctive.*

Mousebirds

FAMILY COLIIDAE

This entirely endemic African family (order: Coliiformes) consists of just six species of very similar-looking frugivorous birds. A problematic group of birds, both to the avian taxonomist and the commercial fruit farmer and gardener. Restricted to sub-Saharan Africa, these small, very long-tailed and crested birds have rather dull brown and greyish plumage, reddish feet and legs, and most have distinctive patches of colour. The blue-naped, red-faced, white-headed, white-backed and red-backed mousebirds clearly indicate by their names what their distinguishing features are.

They take their group name, mousebirds, from the manner in which they scuttle around in the vegetation like tree mice or dormice. The strong bill and large feet aid them in their mainly arboreal way of life. They descend to the ground occasionally, to drink and eat from fallen fruits and berries.

Taxonomists are hard pressed to relate the mousebirds to other bird orders. Some believe they are distantly related to the parrots, others ally them with the turacos on the basis of nest structure, but most accept that they are so different they should stand alone. Their outer toes are reversible and can be used either forwards or backwards.

The mousebirds, sometimes called colies, are highly social and always move around in noisy close-knit flocks. They seldom cover long distances in flight but travel from bush to bush and tree to tree, often flying from a high point in one to a low point in the next. When roosting at night they commonly cluster closely together and hang in a vertical position, clinging to the surface with their strong toes and long, slender claws.

Although known principally for their depredations on commercial fruits, they also eat wild fruits, buds and blossoms. In fact as I am writing this section I know that our newly budding fruit trees are being tackled with gusto by a party of some 20 white-backed mousebirds; when not busy with these they are on the ground feeding on fallen Mexican pepper tree berries.

The flocks break up into pairs in the breeding season. They construct untidy cup nests from a mix of plant material, with a lining of green vegetation that is replaced as it withers and dries out. The female lays between one and five eggs, very occasionally up to seven, and both male and female contribute to incubation and chick care. Just a few days after hatching the chicks leave the nest to perch on nearby twigs and branches but return to the nest at night to be brooded by a parent. In at least some mousebird species cooperative breeding is prominent.

The most studied, the speckled mousebird, may have a breeding group made up of one to three male helpers, which are usually related, and one to three non-related female helpers. The helpers are usually young birds. Of interest is that the dominant male takes a single mate but other males may be polygamous. In the case of this, and possibly other, species the flock's territory is guarded year-round. ■ SEE PAGE 160.

Parrots, parakeets and lovebirds
ORDER PSITTACIFORMES

To many people, particularly in the Western world, the African grey parrot or the colourful macaws epitomise the parrot world. There are some 350 species worldwide (the figure is inexact because of taxonomic uncertainties), but Africa's share of these amazing birds is small: just 22 species. All but the ring-necked parakeet are endemic to Africa and its islands.

The earliest known parrot-like birds were excavated in Lower and Middle Eocene deposits dated to some 40 million years before present; they carry the name *Palaeopsittacus georgei*. Although size and plumage coloration are variable, all are structurally very similar in their overall build, powerful hooked bill and the zygodactyl feet. They have proportionally large heads, short necks and short legs adapted for moving around on tree branches. On the ground they have a characteristic swaggering walk but except at water they usually avoid landing on the ground. Parrots are noisy, especially in roosting trees, and their calls tend to be harsh and grating.

One of the factors that has contributed to their popularity, particularly the African grey, is their ability to mimic human voices and many other daily sounds. Unfortunately, this very popularity has tipped several parrot species towards the abyss of extinction, including a number in Africa.

Some species cover considerable distances between roosting trees and feeding grounds. One of the great movers is the African grey. At roosts it may number in the thousands – the noise levels can be awesome – but they disperse to feed in small groups. They sometimes fly several kilometres over open water, even the sea. Brown-necked parrots are recorded as flying up to 90 km to feeding trees, and red-fronted parrots make round trips that may total 120 km. In some species there are daily and seasonal altitudinal movements.

Most of those small parrots, the lovebirds, have limited distributions but are abundant. Lovebirds take their common name from their habit of sitting close together on a perch and frequently preening each other. They are extremely popular as pets; large numbers are still caught for the cagebird trade, but possibly not as many as formerly. In a period of just four weeks in 1929 some 16 000 black cheeked lovebirds were exported. In the cagebird trade there is an unfortunate trend to breed hybrids to produce mutant colour forms.

Parrots predominantly feed on plant parts; some species are very specialised but others are generalists. The vast majority feed on seeds and fruits, some of which are very hard but are easily handled and cracked with the vice-like bill. Several

Above: *One of the most frequently kept pet parrots is the African grey.*
Left: *The Senegal, or yellow-bellied, parrot is restricted to forest and woodland in West Africa.*
Far left below: *Meyer's parrots live in pairs or small flocks.* (Photo: John Carlyon)
Far left above: *Lilian's, or the Nyasa lovebird.*

species hold the food in one foot, frequently the left, to manipulate it more easily. In some cases feeding parrots discard the fruit pulp and crack the pip to gain access to the kernel, which is then eaten. A number of species are considered to be pests as they readily feast on ripening maize and sorghum crops. Wild grass seeds are also eaten, particularly by the savanna-dwelling lovebirds. Buds, flowers and nectar are recorded as being eaten by a few species and it is likely that all consume these plant parts from time to time.

The vast majority of African parrots nest in tree holes, natural or excavated by others such as woodpeckers and barbets. Some lovebirds also make use of holes in trees, but the red-headed prefers holes in arboreal ant and termite nests, while the rosy-faced nests in rock crevices, under eves of buildings and most commonly in compartments in sociable weaver nests. The females of several lovebird species carry nesting material, primarily grass, tucked among their feathers – a case of flying "haystacks"! Egg numbers are usually two to six but this differs from species to species. Hatchlings are naked but soon develop a covering of down.

Depending on the species it takes two to four years for parrots to reach sexual maturity, but not so long for lovebirds.

Despite the great interest shown in the parrots by people we know remarkably little about many species, particularly in Africa. If one browses through the scientific and popular literature one notices large gaps in our knowledge. For example, the niam-niam parrot of northern Central Africa is an unknown when it comes to breeding habits, as is the yellow-fronted parrot of Ethiopia. Another Ethiopian avian puzzle is the black-winged lovebird; the little we do know of it is based on observations of captives.

On the African mainland the black-cheeked lovebird is considered to be endangered and Fischer's lovebird is hovering on the threshold of being threatened as a result of habitat loss. But the island parrots have suffered the most from human interference. The Mascarene (*Mascarinus mascarinus*) and broad-billed (*Lophopsittacus mauritianus*) parrots of Mauritius disappeared during the second half of the 19th and 17th century respectively. The Seychelles (*Psittacula wardi*) and Newton's (*Psittacula exsul*) parakeets of Rodrigues became extinct at the beginning of the 20th century. The Mauritius parakeet (*Psittacula echo*) is under severe threat in part because of nest-hole raiding by an introduced monkey, the crab-eating macaque (*Macaca fascicularis*). ■ SEE PAGE 159.

THE ORNAMENTAL AND CAGEBIRD TRADE

A wide range of human societies and cultures, past and present, value birds as pets, in particular parrots. The obvious attractions of parrots are their high level of intelligence, potential for tameness, bright plumage and ability to mimic the human voice. Between 1980 and 1992 the legal parrot trade in Africa saw 278 000 Senegal, or yellow-bellied, parrots and 657 000 Fischer's lovebirds entering international pet markets. The illegal trade, which is known to be high, probably saw equal or even higher numbers being smuggled. Many African grey parrots also enter the trade, particularly from rapidly decreasing West African populations. Although captive breeding of a number of African parrot species is successful, production is nowhere high enough to meet market demand.

The main sources of other wild-caught cage and ornamental birds are Senegal in West Africa and Tanzania in East Africa. From the 1950s until at least into the 1980s (more recent figures are unavailable) more than one million birds were exported every year from Senegal. The principal species entering the international trade were or are marabou stork, black crowned crane (a threatened species), chestnut-bellied sandgrouse, various doves and pigeons, parrots, swallows, glossy starlings, weavers, whydahs, queleas, waxbills and mannikins. Of these the waxbills and mannikins often make up almost 50% of all shipments. In this trade, which is legal, a number of bird species are certified as having been caught and shipped from Senegal, although they have never been recorded as occurring in that country! A percentage of trade birds are clearly being caught in neighbouring countries and passed off as originating in Senegal.

The number entering the trade is very high, so the number caught must be astronomical: we know that capture, handling and holding mortality is massive and in some cases, for example swallows, mortality may be as much as 90% of all captures. In recent years it has become increasingly difficult for birds illegally exported, with or without seemingly adequate documentation, to enter Western markets but this is not the case for certain Asian and Arabian markets. We have personal experience of several animal suqs (markets) in the Arabian Gulf states where large numbers of mammals and particularly birds are sold without any international conservation documentation. Birds that frequently feature in large numbers — sometimes horrifyingly so — are both black and grey crowned cranes, demoiselle cranes, greater flamingos, purple gallinules or swamphens, hornbills, ducks and geese, parrots, turacos, francolins, raptors and a whole range of small passerines. We were told of one animal dealer who was willing to buy 4 000 northern pintail ducks! Many of the species in these animal markets have specialised diets and the vast majority die within weeks of purchase. The conditions in which they are kept (we have seen several such establishments) leave much to be desired, to put it mildly. Approximately 100 purple gallinules were bought by a prominent individual and released in a sandy enclosure with nothing other than drinking water and no cover. They shared this small area with some 30 grey crowned cranes, 20 demoiselle cranes and a host of gazelles, fallow deer, plains zebras and wallabies — and this is one of the happier situations! As far as we could establish most of the birds entering the Gulf states do so without documentation.

There is a great need to adequately monitor both the legal and illegal bird trade in Africa. Some species, particularly the seed-eaters, can tolerate a high off-take, but many others are suffering serious declines because of this demand. International controls and regulations are only as good as their enforcement and the officials who implement them.

Sandgrouse

FAMILY PTEROCLIDIDAE

Africa is the stronghold of the pigeon-like sandgrouse, with 12 of the world's 16 species found here. Six of these are endemic and the rest spread eastwards into Arabia and Asia. They are structurally similar to pigeons of the order Columbiformes; some authorities place them together in that order, but in other circles this relationship is not accepted.

Sandgrouse weigh between 150 g and 400 g and all are cryptically coloured, females more so. The short legs and small feet are feathered to the toes; the wings are long and pointed at the tips, and sandgrouse are capable of rapid and direct flight. The belly feathers are structured in such a way that they can hold water for carrying to the chicks, which are often several kilometres away. It is only the male sandgrouse that collects and delivers water to the chicks, which is necessary because sandgrouse live primarily in arid areas: from extreme desert to open savanna or dry mountain country. In spite of their habitat, sandgrouse must drink regularly. Each species is synchronised for a specific time. Some drink in the morning, others in the late afternoon, and others after sunset. Unlike most other bird species, they drink by sucking up water in the bill. When coming to water they may number in the hundreds but they usually disperse on the feeding grounds into small parties. All have characteristic calls which are frequently uttered in flight, and on the ground they make soft "contact" muttering calls akin to the clucking of domestic hens.

Two or three eggs, the normal clutch size, are laid by all species in a simple scrape that is usually not lined or at most has a few bits and pieces from the surroundings. The precocial young soon leave the nest scrape, picking at small seeds with the parent birds in attendance. Very tiny seeds are favoured: one crop, for example, contained an estimated 30 000 seeds!

Arguably the most attractive of the sandgrouse is the pin-tailed, with a range extending from North Africa eastwards into Pakistan and Afghanistan. Although it has a limited African distribution it occurs in the tens of thousands in the high plateaux of the Atlas range, with influxes from the Sahara. The black-bellied sandgrouse has a similar range in the north but also occurs on the Canary Islands. Another abundant species is the chestnut-bellied sandgrouse, with flocks of thousands having been recorded. In Africa it is restricted to the Sahel and dry areas of East Africa, but extends through the Arabian Peninsula and to India. Several other species are restricted to northern Africa but three, the Namaqua, double-banded and Burchell's sandgrouse, are found only in the south. ■ SEE PAGE 158.

Above left: *Pin-tailed sandgrouse males are exceptionally attractive, particularly in full breeding plumage.*
Above right: *Burchell's sandgrouse lands at the water's edge, or on the surface, when coming to drink, unlike the Namaqua sandgrouse which lands several metres away before approaching to drink.*
Left: *Black-faced sandgrouse male. This species generally comes to water one hour after sunrise.*
Far left: *Although the yellow-throated sandgrouse usually lives in pairs or small groups, during migration they may form flocks of up to 1 000 birds.*

Above: *Many races of the green turaco have been described and there is some debate as to whether these should be afforded separate species status.*
Left: *The white-bellied turaco, also known as the go-away bird, favours dry wooded and bush habitats.*
Far left: *The Guinea turaco, like other turaco species, builds a flimsy pigeon-like nest.*

Turacos, go-away birds and plantain-eaters

FAMILY MUSOPHAGIDAE

The name Musophaga means banana/plantain-eater but apparently these birds rarely, if ever, feed on this fruit. Although they are often referred to as loeries or louries in South Africa, we prefer the more international name of turaco. The name "go-away bird" is sometimes applied to the savanna-dwelling turacos and is derived from their frequently uttered call, a curse to the hunter intent on stalking prey, because it alerts the game to the presence of danger.

This is an exclusively sub-Saharan African family of about 23 species (several of the "green turacos" are sometimes lumped as subspecies of *Tauraco persa*). An ancient group of birds, their relationship with other orders is somewhat murky. Turacos are fairly large and the majority are inhabitants of forest and dense woodland. These species have green or blue plumage and red wings. The bright plumage colours of most other birds are produced by the refraction of light from minute structures in the feathers, but turacos have true pigmentation;

the red pigment is known as turacin and the green pigment as turacoverdin. All species have crests, in some more developed than in others. The bills are short and stout, the wings are short and rounded, and the tails are long. They have the ability to bend their outer toes backwards or forwards, an aid to their squirrel-like running along tree branches and through leafy tangles. All are noisy birds; most utter similar croaking calls but the savanna species utter variations on the "go-away" theme. In the case of the forest dwellers the call is often the only indication of their presence. They seldom show themselves, except for the occasional flash of crimson wings as the bird crosses a glade or clearing. They are poor fliers, normally covering only short distances in weak, dipping flight.

The "typical" turacos are those with glossy green or purple-violet plumage and large areas of red on the wing. They are closely restricted to forest and woodland. The greatest diversity of turacos occurs in the tropics. Several have very limited ranges and some are under threat because of habitat destruction. Turacos usually live in pairs or small family parties. Larger numbers of the savanna species, in particular the grey turaco, sometimes gather near water or a good food source. The bulk of their diet is fruits and berries, but seeds, buds, flowers and a range of invertebrates are also taken. The latter includes snails, beetles and caterpillars. However, few turacos have been studied in the wild and little information exists on behaviour, reproduction or diet. The nest is usually a fairly bulky but rather loose platform of sticks constructed in a tree, in which the two, occasionally three, plain eggs are laid. Both parents feed the young with regurgitated food. The young leave the nest long before they can fly, in some cases only 10 days after hatching, crawling along the branches where the adults feed them until they are mature.

The largest member of this family is the magnificent great blue turaco, an inhabitant of tropical forests, where it usually spends most of its time in the canopy. Our first sighting of these superb birds was of a pair that descended to drink at a small crater lake in western Uganda, and it is one of those moments forever imprinted in the memory. They are more gregarious than the other forest turacos and form small parties of up to 18 birds, usually less. Each group has its own territory.

Fischer's turaco is restricted to forest pockets near the East African coast and the island of Unguja, Zanzibar, where it only occurs in the Jozani Forest. It was thought to be extinct there for many years but small numbers still survive. Much more severely restricted and threatened is Bannerman's turaco, occurring in a very limited area of western Cameroon. Prince Ruspoli's turaco is a zoological unknown, even though in its very restricted southern Ethiopian range where it occupies juniper forest it is said to be common. Another species occurs in the Ruwenzoris, taking its name from these mountains, and through Kivu to Mount Kabobo, where it occupies the forested and bamboo-covered slopes up to about 3 400 m.

The three go-away birds, one entirely grey and two with grey and white plumage, occupy woodland savannas, particularly those that have many acacia trees. They are stronger fliers than their forest-dwelling relatives and are often more tolerant of human intruders. The two species of grey plantain-eaters are rather nondescript with black, brown and white plumage contrasting with the stout, yellow bill. ■ SEE PAGE 159.

Above: *The little-known red-crested turaco is restricted to western Central Africa.*
Left: *White-cheeked turacos occupy forests and well-wooded stream banks.*
Far left: *Purple-crested turacos live in pairs or small parties.*

7

Coucals and brood parasites

The coucals are larger than the brood parasites, and they have their own nest, incubate their own clutch and raise their own young in the "normal" fashion. They are believed by one taxonomic school to be closely related to the cuckoos. The yellowbill, an African cuckoo, is not a parasite despite its name. Brood parasites have mastered the art of foisting the incubation and raising of their offspring onto unwitting host species. The advantages include a much larger brood each breeding season than would be possible if the female had to raise her own young in one nest – and she need not waste energy seeking food for hungry chicks.

A verse that we like but whose author we have been unable to identify:

The cuckoo's a bonny bird, he whistles as he flies,
He brings us good tidings, he tells us no lies,
He sucks little birds' eggs to keep his throat clear,
And never sings cuckoo till summer draws near.

We prefer the group name "honeybird", rather than the generally used
"honeyguide", as only one species has been proved to draw humans and honey badgers
to bee hives so that the "Pied Piper" can benefit from the exposed wax and larvae.
Like cuckoos, the honeybird hatchlings do away with the host's legitimate offspring,
thus ensuring that they alone enjoy the fruits of their foster-parents' labour.

Coucals

FAMILY CENTROPODIDAE

Depending on the taxonomic school one follows, Africa is home to seven or nine coucal species, of which only one occurs away from the continent, in south-western Arabia. Coucals are very closely related to cuckoos; in fact they are sometimes only awarded subfamily status and referred to as terrestrial cuckoos. Coucals are medium to large, slow-flying, mainly terrestrial birds. They have rather coarse plumage in black, shades of brown and reddish-brown as well as white, and the feathers on the head, neck and breast are rather bristly. The wings are proportionally short and rounded at the tip, the tail is long, broad and graduated. The heavy bill has a down-curved tip and the feet are well developed.

The coucals have distinctive calls but probably the best known is the so-called "water bottle" call, which resembles the sound of water being poured from a bottle. Unlike cuckoos the coucals are not brood parasites. They construct large, domed nests on or close to the ground, in which the female lays between two and five eggs, the clutch size depending on the species. The nest is built by the male, who also, at least in the

Left: *The Diederik cuckoo is more commonly heard than seen.*
(Photo: Richard du Toit) **Above:** *African cuckoos closely resemble the*
Eurasian cuckoo. (Photo: Richard du Toit)

few species that have been studied, takes responsibility for incubation and most of the feeding of the young.

Coucals eat a wide range of invertebrate and small vertebrate prey, being opportunistic in their foraging. A few species are forest birds, such as the black-throated and Gabon coucals, others occupy dense vegetation in association with swamps, marshes, lakes and riversides. Senegal and white-browed coucals have wide distributions and among the most frequently seen, in the main because of their association with woodland and well-vegetated savannas. They often sit on top of bushes, particularly when they are wet from rain or dew. When alarmed the coucals usually dive for cover in dense vegetation. ■ SEE PAGE 159.

Above: *The Eurasian cuckoo.*
Left: *A number of different races of the white-browed coucal are recognised.*
Opposite page: *Crimson-breasted shrike feeding a black cuckoo chick.*
(Photo: Clem Haagner, ABPL)

Cuckoos

FAMILY CUCULIDAE

The cuckoo, because of its breeding habits, has given the English language the word *cuckold*. The members of this family are often difficult to see and are most frequently identified by their distinctive calls.

Many cuckoos resemble small hawks, with their long wings and tail, and the barring and streaking on the underparts. All are brood parasites, leaving foster parents to raise their young. Each cuckoo selects specific hosts but may parasitise the nests of several luckless species. One of the most remarkable aspects of this host-selection specialisation is that the cuckoo's eggs often closely resemble those of the host. Some hosts are much more adept at detecting the intruder's eggs than others, and if the cuckoo's do not resemble her eggs she would probably throw it, or them, out. What is truly remarkable is that at least in the case of the Eurasian cuckoo a female chooses the host to which she is matched. In this way one species of cuckoo has groups of different females selecting hosts that best suit their egg-matching. Between 12 and 20, occasionally more, eggs may be laid in a season. The female cuckoo lays one egg, occasionally two, in a host's nest, often removing one of the host's eggs, and then moves to the next nest to deposit an egg there. In most cuckoo species a "breeding" territory is known to be established in which all nests of the host species are under the cuckoo's "control".

The second remarkable aspect is that the single, newly hatched cuckoo is programmed to toss the host's eggs, or chicks, out of the nest. This ensures that it will get all the available food, and if it is to grow larger than the host this is important. There are some classic photographs of the diminutive host standing on the back of its "giant" foster chick in order to reach the gaping bill to stuff in food. Once the cuckoo chick is fledged and leaves the host's nest, the foster parents continue to feed it for a considerable time, even though the young cuckoo is quite capable of feeding itself.

A few cuckoo species parasitise the nests of only one or a few host species, for example the great spotted cuckoo, which specialises in crows and jays. In southern Africa the black and white cuckoo (also called the Jacobin) concentrates on bulbuls, and Levaillant's cuckoo parasitises babblers almost exclusively. Both Klaas's and Diederik cuckoos lay their eggs in the nests of many different species, but principally those of small insectivorous passerines.

Insects make up by far the bulk of all cuckoo species' diets, including large quantities of caterpillars, frequently hair-covered and toxic species. We once watched several great spotted cuckoos feasting on large, hairy moth caterpillars with gusto; just the day before I had accidentally brushed against one of these insects and came out in a red, itchy rash. Some species also eat toxic grasshoppers, and very occasionally lizards, birds' eggs and small frogs. A few African species also include a small component of plant food, such as seeds and wild fruits, in their diet.

Several cuckoos are African endemics and all but two, the lesser and Madagascar cuckoos, breed on the continent. Several have extensive African and Eurasian ranges, including the black and white cuckoo, great spotted cuckoo, Eurasian cuckoo and lesser cuckoo. A number of the cuckoos are intra-African migrants, with some undergoing considerable movements.

Most cuckoos lead a largely solitary existence, but some live in pairs and on occasion small parties, especially outside the breeding season. One of the most social during migration is the great spotted cuckoo, with flocks of particularly juveniles on occasion numbering in the hundreds. If one had to choose Africa's most attractive cuckoos, one would have to nominate the small species of the genus *Chrysococcyx*. There are 14 species of which four occur in Africa; two are endemic and the other two nearly so, as they both breed in southern Arabia. The beautiful emerald-green in their plumage, variable in extent between species, separates them from other African cuckoos. Two of the best studied cuckoos belong to this genus, Klaas's and the Diederik, and they are unusual in that both parasitise by and large the same host species.

The yellowbill, or green malkoha, is a large, non-parasitic cuckoo that occupies forest habitats, but it is seldom seen as it skulks through dense vegetation when foraging. ■ SEE PAGE 159.

Honeyguides

FAMILY INDICATORIDAE

With 15 species of honeyguide (more correctly honeybird) endemic to Africa and only two species elsewhere, this continent is obviously their stronghold. Despite their name only two species, the scaly-throated and greater honeyguides, are believed to actively lead, or guide, humans and other mammals to beehives. Hard evidence of guiding is only available for the greater honeyguide.

All are fairly small birds with rather nondescript plumage. The bill varies from species to species. Most, if not all, species include in their diet some wax which they glean from abandoned or active beehives. Apparently the wax is digested by specialised bacteria but further research is required to provide confirmation. Attempts to study these interesting birds are bedevilled by the bird switching to "guiding" behaviour at the approach of the researcher! Insect prey makes up the bulk of their diet, with at least some species supplementing this with fruits, berries and seeds.

The greater honeyguide, with a sub-Saharan distribution, has a highly developed "guiding" system. They are known to guide humans and honey badgers, and possibly other species, to wild beehives. The bird moves from tree to tree in slow deliberate flight, calling frequently and watching the follower's progress. Once the hive is reached it usually waits until the mammal has taken the honey and then it moves in to eat wax. Honeyguides are generally intolerant of each other and are predominantly solitary, or in pairs during the breeding season.

All are brood parasites and the female deposits her eggs in the nests of a wide spectrum of unwitting host species. Most honeyguides and honeybirds parasitise the nests of hole-nesting species such as barbets, woodpeckers, bee-eaters and kingfishers, but those of the genus *Prodotiscus* lay their eggs in open nests. In some species (information is lacking for many others) the female honeybird may puncture or remove the eggs of the host. Newly hatched honeybirds are equipped with a hook on the bill which they use to kill the host's chicks and then push them out of the nest. In this way they ensure, like cuckoos, that they alone will receive all of the food delivered by the foster parents.

Several species have fairly wide but often disjointed distributions in the tropics, but some others have very limited ranges. In the latter category is the yellow-footed honeyguide, which has only been recorded in two small areas in Liberia and southwestern Cameroon, but virtually nothing is known about this bird. ■ SEE PAGE 161.

Above: *A lesser honeyguide about to leave the nest of a pied barbet, the host species that reared it.* (Photo: Peter Steyn)
Left: *The greater honeyguide takes its name from the fact that it guides people to bees' nests.* (Photo: John Carlyon)

Above: *The shaft-tailed whydah male.* (Photo: Brendan Ryan, ABPL)
Left: *Pin-tailed whydah male, with its long tail, and female.*
Far left and right: *Eastern paradise whydah male in full breeding splendour.*

Whydahs and widowfinches
FAMILY VIDUIDAE

This interesting group of small seed-eaters parasitise members of the Estrildidae family and certain warblers. Some of these brood parasites are very specific in their choice of unwilling host, others lay their eggs in the nests of several host species. Male whydahs and widowfinches are polygamous and their harem members lay one or more eggs in each host nest, often removing the host's eggs.

There are 19 species in Africa. The males of several in breeding plumage are impressive indeed, but once the performing and mating is over then they become drab little fellows just like the females. The whydah males grow special tail feathers: long and broad in some species (such as the paradise whydah), fine in others (such as the straw-tailed whydah).

The tiny indigobirds, or widowfinches, are rather uninspiring little chaps clad in mainly black plumage. The females are nondescript brownish birds. ■ SEE PAGE 168.

8

Aerial feeders

The mixed grouping of unrelated birds covered in this chapter share the habit of taking the bulk of their insect prey on the wing. All are adept at aerial manoeuvring, reaching its highest form in the swifts. Varying degrees of anatomical adaptations aid them in their foraging. With the exception of the nocturnal nightjars, all species are active in the daytime. During the course of a typical summer's day I can look out of my office window and watch wheeling and screaming groups of little and white-rumped swifts, European bee-eaters, barn and greater striped swallows – all aerial masters. As the sun sets and most have taken to their roosts, the fiery-necked, rufous-cheeked and freckled nightjars begin their various "songs", a prelude to their nocturnal airborne foraging.

Most of the species in this grouping eat their prey while on the wing but many of the bee-eaters carry larger, and particularly stinging, prey to a perch to be beaten and broken before being swallowed. When an abundance of prey, such as termites, is available on the ground many species will land to gorge themselves on this bounty. We have observed several greater striped swallows clustered around a hole from which were emerging ant alates, snatching them up before they could take flight, sharing this feast with a Karoo toad.

Bee-eaters

FAMILY MEROPIDAE

Africa has 19 species of these avian jewels, of which 15 are endemic. Only six other species occur elsewhere. All are relatively small, lightly built aerial masters, with colourful plumage, long and slightly down-curved bills and pointed, scythe-shaped wings. In most species the two central tail feathers extend well beyond the rest to form a point, but a few have squared-off tails, and one a deeply forked tail. Their legs and feet are weak and they seldom walk on the ground.

Left: *The little bee-eater often perches on a low branch from where it hawks insects.*
Above: *Pearl-breasted swallow.* (Photo: Lanz von Horsten, ABPL)

A fairly general division can be made between the smaller and larger species, in that the former tend to be resident and non-migratory, occurring singly, in pairs or very small parties, whereas the larger are usually in flocks and are more prone to short and long-distance movements. The small species tend to snatch their insect prey from a favoured perch close to the ground and take the victim back to the perch to consume it there, whereas the larger species mostly hunt in aerial hawking flights. Although all species take a wide variety of mainly flying insects, bees and wasps, as the birds' name implies, are an important component of their diet. Stinging insects are rubbed vigorously on the perch in order to get rid of the venom.

Although most species are associated with savanna and woodland, a few are restricted to the tropical forest belt. These include the black-headed, blue-headed and rosy bee-eaters. The two species with the most restricted ranges are the cinnamon-chested (found only in the highlands of East Africa) and the Somali (only in the Horn of Africa and northern Kenya).

Among the colonial nesters the rosy bee-eater takes the prize for high-density housing. One colony of nesting burrows numbered no less than 23 700; by comparison with these cities the colonies of the two carmine bee-eaters are but towns with 100 to 1 000 burrows, although some may boast 10 000. Most colonies contain far fewer burrows which are well apart from each other, but in some species the holes are closely clustered. At some carmine bee-eater colonies the density may reach 60 entrance-holes per square metre. This presents one of Africa's greatest avian sights: the brilliantly coloured birds wheeling and calling, and perching at the holes in the earth bank, converting the dull soil into a shimmering swirl of colour. A few bee-eater species excavate their burrows in nearly level ground where suitable banks and cliffs are not available. We have encountered burrows of the little green bee-eater in silt mounds no higher than 20 cm above the surrounding soft sand. In the case of the rosy bee-eater, some colonies are located on flat or slightly sloping sand substrate with a scattering of low vegetation. Burrows are excavated with the bee-eater's feet while the wings and bill are used as supports. Both members of a pair help with the excavations.

Research on the red-throated and white-fronted bee-eaters has been intensive and they have been found to have among the most complex societies known in the bird world. A fair number of pairs, usually about a third in a colony, are assisted in burrow excavation, incubation and raising the young by "helpers"; up to three and even five have been recorded. Although the entrance to the burrow is controlled by the parents and helpers, a few but not all colony members are permitted entry for a short time. In the case of the white-fronted, and possibly the red-throated, a social clan system operates, each clan consisting of between one and five breeding pairs and their helpers. Each clan also defends the feeding territory from other clans and individuals, even though this may be a considerable distance from the nesting colony. ■ SEE PAGE 160.

Far left: *Little green bee-eaters commonly associate with cultivated areas, where they find abundant prey.*
Left: *The southern carmine bee-eater lacks the blue-green colouring below the eye-stripe.*
Below right: *Red-throated bee-eaters occupy areas of bushed grassland, often along streams.*
Below middle: *Swallow-tailed bee-eater fluffed up, having just regurgitated a pellet of undigested prey remains.*
Below, far left: *White-fronted bee-eaters forage singly, or in small parties, but roost at night in large flocks.*

Above: *Freckled nightjars, like other members of the family, are so well camouflaged that they are difficult to spot at their daytime roosts.*
(Photo: John Carlyon)
Left: *No attempt is made by the rufous-cheeked, or any other nightjar, to construct a nest.*
Far left: *Like all nightjars, the fiery-necked is active only at night.*
(Photo: John Carlyon)

Nightjars

FAMILY CAPRIMULGIDAE

When the sun has set, Africa's 25 species of nightjar begin to stir, preparing for a night of hunting insects on the wing. In general the African species are poorly known and this is aggravated by the fact that many look almost identical. All species have cryptically marked plumage and unless moving they are extremely difficult to see. When resting up during the day they lie on dead vegetation, rocks, gravel or branches, improving their already superb camouflage by drawing the large eyes to merely a slit. These small to medium-sized birds all have proportionally long, slender and often pointed wings. Their flight is strong and silent, and although it may be rapid and direct, it is reminiscent of the wing-strokes of large nocturnal emperor moths. The large eyes are an adaptation to night hunting and the wide bill gape is an aid to taking insects in flight. This massive gape is further enhanced in most species by long facial bristles (called rictal bristles) that increase the effective area of their "flytrap" mouths. The legs are short and the feet small, with a comb-like structure down the outer edge of the middle claws which is presumed to serve a grooming function.

Nightjars do not construct a nest but lay the eggs, usually two, on bare ground or rock. Little is recorded about the nestling period but in some species the chicks move to different sites before they are fully fledged and able to fly. Presumably this is a means of avoiding the attentions of predators.

A few nightjars, such as the long-tailed, standard-winged and pennant-winged, are fairly easy to identify, provided in the case of the last two mentioned that one is looking at a breed-ing male because only they carry the distinctive appendages! Further confusion is sown by the fact that in some species the patches on the wings and tail are white in the males but brown in the females. Their calls, however, are distinctive and provide the best clue to their identity. Listening on a still early summer's evening in the upper Karoo to the liquid "good lord deliver us" call of the fiery-necked nightjar and the inelegant churring of the rufous-cheeked nightjar would lead one to believe that identifying nightjars is simple, but in areas where several species occur together, even the calls can become confusing!

In Africa, one or more nightjar species are found in virtually every habitat, from the Bates's and brown nightjars of the tropical lowland rainforests, to the swamp nightjars inhabiting a range of moist habitats, to mountain inhabitants such as the montane and Ruwenzori nightjars. Many occupy savanna and arid areas, but the greatest diversity of nightjar species is reached in the tropics with its year-round abundance of insect prey. Of the African nightjar species 21 occur only in the tropics, have at least part of their range in the tropics or migrate there from colder regions. The rufous-cheeked nightjar breeds mainly in southern Africa but with the approach of the winter it migrates northwards, and the European nightjar then crosses through much of Africa to as far south as Cape Town. When the southern winter begins the European nightjars have already returned to their northern breeding grounds. Unlike most African nightjars, which are ground-roosters, the European representative rests on tree branches. ■ SEE PAGE 159.

Rollers

FAMILY CORACIIDAE

The rollers, of which eight species occur in Africa, are pigeon-sized birds that are notable for their beautiful plumage dominated by blues, and their dramatic aerobatics. In typical rollers it involves a rapid tumbling display flight, in the broad-billed and blue-throated rollers it involves sequences of steep swoops. The combination of the bright colours and impressive aerial displays is not backed up by a good voice. Their raucous, loud squawks and screeches are often repeated, although when perched they are quieter.

The sexes are alike. They have large heads and the bill is fairly short but stout, particularly in the two broad-billed rollers. They have long, broad wings, short legs and strong feet. In most species the tail is not particularly long but several have extended, pointed streamer-like outer feathers, and in the racket-tailed roller the tips of the streamers are broadened. Like kingfishers and bee-eaters, nestling rollers go through what is delightfully known as the "hedgehog" stage: they retain the sheaths around their growing body feathers until shortly before leaving the nest.

The ground-rollers (Brachypteraciidae) of Madagascar and the single cuckoo-roller (Leptosomatidae) of the Comoro Islands are believed to be closely related to the African continental rollers.

Rollers are primarily solitary or in pairs, but loose groupings may be seen in some species during migration. However, the blue-bellied roller commonly consorts in small groups and they often gather in loose feeding flocks, a special sight indeed as these aerial blossoms lunge and swirl in pursuit of prey. European rollers are usually solitary or in pairs but in the spring along the East African coast migrants gather in the tens of thousands. Crossing ocean and desert, some of these rollers cover as much as 10 000 km between their summer and wintering grounds.

Feeding strategies can be divided into two groups: only the broad-billed rollers take their insect prey on the wing, the rest are "perch-and-pounce" specialists. These usually have favoured hunting perches from which they watch the ground for insects, other invertebrates, lizards and small snakes. The lilac-breasted roller has been recorded taking small birds, and it is probable that other species include such prey items in their diet from time to time. Rollers are great fans of savanna fires and will congregate and snatch up small animals trying to flee the inferno. Rollers are cavity-nesters, usually in tree holes, but some species not infrequently use suitable sites in buildings or occasionally in termite mounds, crevices and holes in cliffs and earth banks. ■ SEE PAGE 160.

Far left: *The lilac-breasted roller commonly hawks insects and other prey ahead of grass and bush fires.*
Left: *The purple roller, like most other rollers, has a dramatic aerial tumbling display.*
(Photo: John Carlyon)
Below: *Roller sun-basking, with interested observer!*

Left: *Because of taxonomic confusion the exact range of the pale crag martin is not known.*
Above: *Red-breasted swallow.* (Photo: Brendan Ryan, ABPL)

Swallows and martins
FAMILY HIRUNDINIDAE

Swallows appear frequently in poetry and prose, plays and fable. It was believed in Europe in centuries past that as autumn approached and the barn swallows gathered, soon to disappear, these birds burrowed into river mud to await the arrival of spring. Today of course we know that they migrate southwards into Africa but when people knew nought of these travels, they had to find some conceivable explanation.

The swallows and martins are a highly specialised group of birds adapted for aerial hunting of insect prey. They spend much of their day on the wing, but not to the extent of the totally unrelated swifts. The superficial similarities to swifts are a classic case of convergent evolution at work. All are small with long, narrow and pointed wings, and in most cases with a backwards sweep, the tail is square, shallowly or deeply forked and with or without long outer feathers. Many have glossy metallic blue-black upperparts, but some have dull brown or grey plumage. Like nightjars and swifts, the swallows and martins have tiny bills but an enormous gape enhanced in some species by short, sharp rictal bristles to even further enlarge the "flytrap". The legs are short and the feet tiny, although they perch with ease.

The family as a whole has a very wide distribution, being absent only from the polar regions and some oceanic islands; 47 species occur on the African continent, a large percentage of the world total of approximately 74 species. Of the African species more than half are endemic. A few are more or less sedentary, some are nomadic and others undertake notable migrations, two of the distance record holders being the barn swallow and the northern house martin. There have been recent additions to the African list in the form of the Ethiopian and Red Sea swallows.

To watch thousands of barn swallows gathering to roost in reed-beds in the evening, or hundreds of sand martins hawking insects low over a river or lake edge, to observe the building skills of these small birds that can construct a simple cup of clay pellets, or a sealed bowl with or without an attached tunnel, or a tunnel in the sand – one can only marvel at their grace and accomplishments. Many also have soft twittering calls that are not unpleasing to the human ear.

An unusual member of the family is the African river martin, so much so that it is placed in a separate subfamily which it shares with a single Asian relative. They have glossy black plumage, a stout (for a swallow) yellow bill and a fairly short, rounded tail. They occur along the Congo River and other suitable large rivers, lakes and estuaries in the western Congo basin. Strongly flocking birds, they occur in hundreds together, thousands when migrating within the Congo Basin. They nest in burrows which they excavate in flat sandy ground, often hundreds close to one another. Other species restricted to the Congo basin include Brazza's martin and the Congo martin, both colonial nesters. The swallow known to have the most restricted range in Africa is the white-tailed swallow, occurring in a small area in south-western Ethiopia.

Several swallows and martins are solitary breeders, in burrows or nests, but many are colonial and some spectacularly so. Preuss's swallow, a mainly West African species, usually nests in colonies of several hundred pairs but some may contain up to 5 000 birds. These birds also construct one of this family's greatest architectural marvels. It is a flask-shaped chamber with a long ground-pointing tunnel constructed of mud pellets. The nests are often so closely clustered that they almost seem to be a single unit. The South African swallow also lives in densely packed nesting colonies and its nest has a similar construction but lacks the long entrance tunnel. In these and many other swallow species pairs usually return to the same nest each year, for as long as it withstands the rigours of nature. If damaged the nest is repaired or if completely destroyed a new one is built. The nests of the greater and lesser striped swallows are very similar to those of Preuss's swallow but differ in that the entrance tunnel is horizontal and the nest is attached to a rock wall, building or tree. Some species construct cup-shaped structures that may be closed or open, while others favour half cups made up entirely of mud pellets or a mixture of mud and plant material.

We have observed both greater striped swallows and rock martins constructing their flask and open half-cup nests on several occasions and it never ceases to amaze us. Backwards and forwards they fly carrying the small mud pellets for nest construction, making many hundreds of journeys. South African swallows apparently require between 1 300 and 1 800 pellets for each nest; the northern house martin collects an average of just over 1 000 pellets. Those species that build the more elaborate nests may take several weeks to complete them.

Although the members of this family are principally aerial hawkers, taking their insect prey on the wing, it is not unusual for some to feed on the ground. ■ SEE PAGE 161.

Above: *Pearl-breasted swallow feeding young.* (Photo: Warwick Tarboton, ABPL)

Swifts and spinetails
FAMILY APODIDAE

Many of the African swifts are endemic or near endemic to the continent or associated islands. Most have wide African ranges but a few are localised; the São Tomé spinetail is restricted to the island of that name, as well as Principé in the Gulf of Guinea.

The black spinetail is known from a few isolated locations scattered across the tropical forest belt but given the very difficult nature of this habitat (for humans!) it probably occurs continuously but is merely overlooked. Many swifts and spinetails are notoriously difficult to identify in the field and in such habitats can be doubly so. Schouteden's swift is known from just a few specimens collected in the far eastern Democratic Republic of Congo. In fact, even some of the common swift and spinetail species have very patchy distributions, if one believes the published range maps. Although some species are certainly rare and localised it is likely that others are much more widespread but many areas lack observers, or more particularly those with the skills to identify these avian masters of flight.

These are proportionally the fastest flying birds on earth, with speeds of over 150 km/h having been recorded in some species. The name of the family implies that they lack feet, which is of course not true, but their walking skills are minimal. In the evolution of their superb flying skills, their long, slender, scythe-shaped wings developed at the expense of their tiny legs and feet. The swifts are the most aerial of birds, eating, drinking, gathering nest material, copulating and in some species even sleeping on the wing. Apart from the distinctive wing shape and tiny feet, they have large heads and a small bill

Above left: *Greater striped swallow.* (Photo: Warwick Tarboton, ABPL)
Above: *White throated swallows.* (Photo: Hein von Horsten, ABPL)
Left: *Little swift, constructing its nest of feathers, vegetation and saliva.* (Photo: Brendan Ryan, ABPL)

but with a wide gape which resembles that of the nightjars but lacks the bristle fringe. In some species the tail is deeply forked, in others short and squared off, and spinetails have a row of short, straight and spiny projections extending beyond the tail, serving as an aid in their vertical roosting positions. The plumage in most species is brown or black, and a number of species have white on the throat, rump or belly. Because of their tiny feet they never alight on branches and only very rarely on the ground, because they can neither perch nor easily become airborne again from the ground. Instead they cling to a vertical surface which enables them to "launch" safely.

Most are quite vocal, particularly when wheeling and circling in flocks. The little swifts nesting in our home village fre-

quently consort in tightly bunched, wheeling and screaming flocks of 20 and more individuals in the early mornings and late afternoons. These boisterous aggregations are sometimes referred to as "circuses" and as the "screaming display". During the spring and summer they often join with smaller numbers of white-rumped swifts in these aerial acrobatics. We have often watched flocks of pallid swifts wheeling and screaming over nesting sites in buildings with many crevices and holes in the walls, and the skill with which they zip over, above and around each other cannot help but remind one of the amateurish (in comparison) aerial skills of "Top-Gun" human fighter pilots.

Swifts are also unusual, and varied, in their nest-building. However, for several swifts our knowledge of behaviour and nesting is minimal. Many construct their small egg platforms on rock overhangs, in cave entrances and in crevices in either rocks or building. Some species are solitary nesters, some form small colonies and still others breed in colonies including dozens of pairs. The little swift constructs a fairly bulky closed "bag" of plant material and feathers held together by saliva, often with several nests touching. White-rumped swifts will appropriate the nests of little swifts as well as some swallows and other bird species. The horus swift has evolved its own nesting technique, appropriating the burrows of ground-nesting birds such as the bee-eaters, kingfishers and sand martins.

Both parents take responsibility for nest building, incubation and the raising of the young. The eggs are white and are low in number, normally one or two but in some species as many as four to six, and after hatching the chicks remain in the nest remarkably long for such small birds. Although our knowledge of most species is scanty, a pair bond may last for several years in at least a few species; this is best documented in the alpine swift. It also seems likely that in at least a few species the same nesting site is used year after year.

Saliva, surprisingly, plays a critical role in holding the nesting material of swifts together, and the saliva glands of nest-building birds enlarge during that time. Within the swift family there are some remarkable nest structures. Probably the strangest of all is that of the African palm swift: it glues a small, flat and rimless pad of feathers to the inner, hanging leaf or underside of a palm frond. The female then glues her clutch of one or two eggs to the pad with saliva, with the parents alternating to incubate the eggs in a vertical position, holding on with the toes. On hatching the naked chicks hook themselves in place on the pad and there they stay until they are capable of flight. Several swifts construct open cups of plant material, grass or feathers, glued together and to the rock or tree surface with saliva. The spinetails build their nests mainly in tree cavities and hollows, baobabs being the favourite of at least one species. The bat-like spinetail nests in old mine shafts but it is likely that tree holes are also used. ■ SEE PAGE 159.

9

Hole nesters

The only thing that birds in this diverse collection have in common is their preferred nesting sites: self-excavated or natural cavities. The woodpeckers and barbets are the master "drillers", excavating impressive nesting chambers in tree trunks. Many other birds appropriate these holes, or make use of natural cavities. Some kingfishers excavate deep burrows in earth banks. Certain starlings also make use of natural tree holes, and the pied starling nests in earth banks, but for the sake of convenience they are described in chapter 11.

With the exception of the ground hornbills, the other hornbills occupy natural tree holes into which the female is sealed while she incubates the eggs and cares for the young in the early stages of development. Many species covered in this chapter are African endemics, a few others spill into northern Africa at the edge of their Palaearctic range. These birds are diverse in their hunting strategies: woodpeckers obtain their insect food mainly from trees; kingfishers may be fish-hunters, plunging into the water to snatch up their piscine prey, or foragers on dry land. Some birds hawk their prey from a perch, others work through trees and bushes picking off tasty morsels, still others walk along the ground taking any edible animal food that can be swallowed. Particularly in the case of many forest species, our knowledge is lacking.

Barbets and tinkerbirds

FAMILY CAPITONIDAE

These are mostly tiny to smallish, plump birds with short tails (except in the genus *Trachyphonus*), short necks and large heads armed with heavy bills. The majority of species have bright colours or bold markings. They also have a bristle cluster around the base of the bill, and this feature has given them their name of barbet or tinker barbet.

Although also occurring outside the continent, this family is best represented in Africa, with 44 species (some authorities

believe there to be two or three fewer). Most species have rather harsh and monotonous calls, often repeated, and because of this they are sometimes nicknamed "brain-fever birds". The smaller species are mainly solitary or live in pairs, but some of the larger barbets have a social life style, living in groups that include nest-helpers. The naked-faced barbet is highly social, nesting and roosting in large numbers, with up to 250 pairs having been recorded in one tree.

Barbets are perhaps not as efficient at excavating feeding and nesting holes as woodpeckers, but nevertheless with those powerful bills they can really make the woodchips fly. They

Above: *D'Arnaud's barbet is a colourful, active and noisy bird of savanna woodland.* **Left:** *Several races of the golden-tailed woodpecker are recognised.* (Photo: Daryl Balfour, ABPL)

usually nest in trees but on occasion in termite mounds and in earth banks. Apart from using the holes for egg-laying and the raising of young, the adults roost in them, often throughout the year. Many holes are located on the underside of a branch and in this way they are protected from rain. Because of their hole-nesting little is known about such factors as incubation or fledgling periods in many species. Two to four white eggs are the usual clutch and in those species studied both parents feed the young, sometimes with the aid of helpers. Nest cavities are kept clean, with the adults carrying the droppings of the young well away. Most African barbets are mixed feeders, taking fruits and insects, but young are fed mainly on invertebrate food.

Food is gleaned from vegetation; some species more frequently forage on the ground, while others feed in the air (particularly the smaller species).

As far as habitat is concerned, barbets can be broadly divided between the forest dwellers and those of woodland savanna mosaics. Within their chosen habitats they can be widespread and common but a few species have very restricted ranges. Red-faced barbets occupy a tiny area of woodland just to the west of Lake Victoria, and Chaplin's barbet extends only marginally outside the Kafue basin in Zambia. There are also those that have strangely disjunct distributions: the green barbet, western green tinkerbird and white-headed barbet. ■ SEE PAGE 160.

Left: *Red and yellow barbet, Tanzania.* (Photo: Clem Haagner, ABPL)
Above left: *The white-eared barbet roosts communally in tree holes at night.* (Photo: John Carlyon)
Above right: *The African hoopoe, sometimes considered to be a subspecies of the Eurasian hoopoe, is very widely distributed. Its flight is butterfly-like.*

Hoopoes

FAMILY UPUPIDAE

There is considerable argument as to whether two species should be recognised, the Eurasian and African, or just one known as the hoopoe. Nine races are recognised throughout the extensive Eurasian and African range. Some populations are resident, others are local and long-distance migrants. It is generally believed that few African populations are totally sedentary, but in a few areas they are largely so. In our home village at least two pairs remain in the vicinity throughout the year with a slight influx of additional birds in spring.

The hoopoe is a distinctive bird with an impressive erectile crest, broad, rounded and beautifully patterned wings and a flight action that brings to mind the flapping but strong flight of a large butterfly. The bill is fairly long, pointed and down-curved and the feet are well developed. Most foraging is done on the ground. The very distinctive and far-carrying "hoo-poo" call gives this bird its onomatopoeic common name.

Hoopoes occupy a wide range of habitats but avoid dense forest. They frequently associate with human habitats such as fields and gardens, and not infrequently nest in human struc-

tures. Long ago already these handsome birds obviously made an important impression: they feature in the wall paintings of ancient Crete and Egypt.

The long bill is used for probing in the soil for insects and their larvae as well as a whole host of other invertebrates. They often dig vigorously with the bill, and also use it to flip over pieces of vegetation, wood and animal droppings. On occasion they will take small lizards, snakes and frogs. Larger prey is thoroughly beaten and "tenderised" before it is swallowed. We have watched hoopoes hawking butterflies quite successfully but they are principally ground-feeders.

It is in their nesting habits that hoopoes fail to live up to their handsome appearance. Nestlings defecate in the nest and the accumulation of this and regurgitated pellets, plus a foul-smelling secretion from the preen-gland, emit a truly notable "perfume" that is not likely to be overlooked by anyone in proximity to a hoopoe nest. The nest, in tree holes, termite mounds, old walls or crevices, normally has a narrow entrance and is barely lined with a few grass blades. ■ SEE PAGE 160.

Hornbills and ground hornbills

FAMILIES BUCEROTIDAE AND BUCORVIDAE

This is one of our favourite groups of birds, certainly not beautiful but they really have style! Anyone who has watched ground hornbills pacing across the savanna, a yellow-billed hornbill trying to cadge a free meal at a game reserve campsite, or a male hornbill delicately feeding his mate ensconced in her temporary tree-hole prison, has to agree.

The 25 African species fall within just two families and three genera, only one of which is not endemic. This is the African grey hornbill, sometimes referred to as the little grey, which extends into south-western Arabia. Hornbills range considerably in size, from the black dwarf (rarely as much as 130 g) to the heavyweight southern ground hornbill (6 kg or more). All have large bills. Several species have the bill surmounted with a casque, which may be a simple low structure or prominent and bulky, those of males always being larger. Only a few bird species have eyelashes but those of the hornbills are impressive, something of which any Las Vegas showgirl would be proud. The broad wings carry them in dipping flight; short periods of strong flapping are followed by a longer downward glide.

Hornbills differ from all other birds in having the vertebral axis and atlas fused. Their nesting habits are also unique in that the female incubates her eggs and raises the chicks in the early stages in a sealed cavity where she remains up to the time when it is necessary to help her mate gather food for the growing youngsters. She then breaks out and in most species the youngsters reseal the nest entrance mainly with their own faeces.

Most plumage comes in combinations of black and white, and greys and browns. The bills of some of the smaller species are yellow, red or both. Colourful bare facial and throat skin is most developed in the black-casqued wattled hornbill and the two ground hornbills. In a few species the sexes are similar but in several of the smaller species they differ in bill colour. The bill and casque are usually hollow or honeycombed, and despite their size are lightweight. The seemingly all-black ground hornbills reveal large white "windows" in flight. Most species have fairly long tails but that of the white-crested hornbill is exceptionally long, as is its curly topknot. Except for the ground hornbills, all have short legs but the feet are well developed.

Watching ground hornbills stride across the open savanna always, for us at least, calls to mind a small group of clergy out

Far left: *The black and white casqued hornbill is a species of forest and dense woodland.*
Above left: *Southern yellow-billed hornbills are common in a number of woodland savanna types.*
Above right: *The trumpeter hornbill is a gregarious, noisy bird with a range of squealing and wailing calls.*
Left: *Southern ground hornbills have a distinctive, far-carrying call.*

to solve the world's problems. These two species, the southern and Abyssinian, are the largest of the hornbills. Apart from their predominantly terrestrial existence, they differ from other hornbills in that they do not seal the nest cavity entrance, nor do they bother to keep the chamber clean. They also primarily eat animal food and are the only African representatives to engage in cooperative breeding.

We have frequently observed the southern ground hornbill but our first sighting of its relative the Abyssinian was just a few years ago at the foot of a cliff at the Murchison Falls in Uganda. Here, against the roaring backdrop of one of Africa's great waterfalls, we had the privilege of watching these fine birds foraging. In contrast to the southern ground hornbill, the Abyssinian is little known. The young develop adult plumage coloration more rapidly than the southern, and from what little is known they seem to have a more varied diet, eating more vegetable matter. Neither species occupies forest or dense woodland but they enter the edges where they may roost and nest.

The calls are characteristic and cannot be mistaken for those of other species, the southern uttering a far-carrying, booming "hoo-hoo-hoo-hoo". Prey may be picked from the ground or vegetation as the bird strolls along, but they also frequently dig out their mainly animal food. They also commonly break up animal dung, particularly that of elephant, seeking insects. Cooperative breeding groups consist of between two and eight birds, usually with only one adult female. Each breeding pair and the associated helpers defend a large territory, up to 100 km² found in one study. A pair may use the same nesting site, in a natural tree hole, cliff or earth bank, for several breeding seasons. The female lays one or two eggs; in the latter case the second chick does not survive but starves to death. While the female is incubating the eggs she is fed by other members of her group. This continues for the full 40 days up to hatching, and she remains at the nest for several days longer. Unlike the other hornbill species which keep their nest cavities clean by squirting their faeces, with amazing accuracy, through the narrow slit of the sealed entrance, the young ground hornbill is a messy little so-and-so that defecates on his "bedding".

The hornbills of the genus *Tockus* are most commonly seen, as many species occupy open woodland and savanna associations. Any visitor to savanna conservation areas is sure to encounter several species such as the red-billed, the two yellow-billed or African grey hornbills. A few extend into fairly dry areas but only Monteiro's hornbill occupies true desert, in the Namib. Some species form monogamous pairs

Above: *One of the most widely distributed woodland savanna species is the grey hornbill; there is also a population in south-western Arabia.*

which hold territories throughout the year, others appear to defend territories only during the breeding season. In the latter category falls the red-billed hornbill, which may form flocks of dozens of birds during the dry season. Although direct conflict between these birds does occur, they make great efforts at peaceable communication with combinations of calling and displays from a prominent perch. The calls are mostly a variation on a "kok-kok-kok-kokok ..." theme and are typical sounds of the African woodlands. Displays are variable and include wing spreading, head and body nodding, "sky-pointing" and tail wagging, nearly always in tandem with the calling.

The generally omnivorous diet of hornbills can vary seasonally and according to species. Some take more animal food, others have fruits and other plant material dominating their diet. Although the animal component is made up mainly of insects and other invertebrates, they will not hesitate to snatch up lizards, frogs, young birds and mice if the opportunity presents itself. We have watched several species of hornbill following in close attendance to grazing and browsing mammals, snatching up small animals disturbed by their hoofs. Von der Decken's hornbill is said to have a special relationship with troops of Africa's smallest carnivore, the dwarf mongoose. The birds pick up insects that escape the attention of the mongooses, and the latter are said to benefit from the hornbills' alertness, which acts as an avian predator warning system. In its limited range Bradfield's hornbill spends a lot of time foraging on ungulate droppings, particularly where there are large accumulations. In all species the food item is taken in a somewhat dainty manner in the tip of the bill and delicately "thrown" into the throat. Observing black and white casqued hornbills in western Uganda feeding on small wild figs, we were amazed at how they plucked each fruit and "tossed it back", not losing a single fig in the process.

A few of the *Tockus* hornbills are forest dwellers, as are the eight members of the genus *Ceratogymna*. Some occur across the tropical forest belt, some show a strong preference for montane forest and others for riparian forest. The forest *Tockus* principally eat insects and other animal life but the larger *Ceratogymna* hornbills consume mainly fruits and berries. Silvery-cheeked hornbills have been recorded as attacking and eating fruit bats at their tree roosts. With the exception of the piping hornbill, members of this genus are medium-sized to large. They are noisy and fly with powerful and clearly audible wing-strokes. Several species may be observed as singles, pairs or small parties, but not infrequently large flocks may be seen. Trumpeter hornbills often forage in parties of 30 or more, and at the roosting tree or trees 200 or more individuals may gather. Silvery-cheeked and black and white casqued hornbills may also forage and roost in substantial numbers. In the forests of Mount Meru in Tanzania, among other places, both silvery-cheeked and trumpeter hornbills are common. ■ SEE PAGE 160.

Kingfishers

FAMILIES ALCEDINIDAE AND HALCYONIDAE

Africa's kingfishers can be divided into two broad groupings, the water-associated species and the dryland foragers. There are invertebrate-hunting species, those that mix invertebrate and fish hunting, and the true fishers.

A charming tale about the common kingfisher, stemming from the mythology of the Bible, has it that this then dull grey bird left Noah's Ark and flew straight towards the sun; in the process its breast feathers were scorched red and its back reflected the colour of the evening sky. An imaginative tale for a most beautiful bird. Another delightful, albeit tragic, tale is from Greek mythology: Alcyone married Ceyx, the son of Hesperus; on learning of his death in a shipwreck she committed suicide by drowning and the gods felt such sympathy for this loving couple that they turned them both into kingfishers.

Worldwide no less than 87 kingfisher species are recognised, ranging in size from the minute (10 g) dwarf and pygmy king-fishers to the Australian kookaburras (500 g). In Africa the largest and heaviest species (approximately 350 g) is the appropriately named giant kingfisher. It is generally believed that kingfishers had their origins in Asia, and only one genus and 14 species are endemic to the African continent. Some species are migratory, others have a resident and migratory component.

Kingfishers are all stocky, short-necked birds with large heads and long, powerful, pointed bills. Bill structure differs between the land and water kingfishers in that the bill is dorsoventrally flattened in the dryland species and laterally flattened in the fish-eaters. Most African species are brightly coloured, such as the malachite and pygmy; others are distinctively marked in black and white, such as the pied kingfisher. Their magnificent plumage varies little, if at all, between seasons, and the sexes are the same in coloration but with some

Far left: *The brightly coloured malachite kingfisher nests at the end of a tunnel about 1 m long, with a chamber lined with undigested prey remains.* (Photo: John Carlyon)
Left. *In Africa the white-collared kingfisher is only known from the lower Red Sea coastline.*
Below: *The woodland kingfisher, as its name implies, is a dryland hunter of insects, lizards, snakes and frogs.* (Photo: John Carlyon)
Below left: *The striped kingfisher is a dryland species, nesting in tree holes.*

Right: *In spite of its mainly insect diet, the brown-hooded kingfisher has been recorded catching small birds.*
Far right: *A widely distributed species, the pied kingfisher occurs through much of Africa, the Middle East and eastwards to China.* (Photo: John Carlyon)

subtle differences in a few species. All are fast and direct fliers, and anyone who has observed for example a malachite or common kingfisher in flight will have a visual image of a jewel-like blur. As is the case with many "jewel" birds the voice rarely lives up to the picture, and the kingfishers have rather harsh, screeching calls. It is often these calls that indicate their presence, and they can be a useful way of establishing numbers along waterways and lakes. We used this method to do a rough count of white-collared kingfishers in a mangrove thicket on the Gulf of Oman.

The "typical" kingfishers – those that live up to the human image of the plunge fishers – include Africa's largest and smallest species. Some are fairly easy to observe, such as the pied kingfisher, but riverine forest dwellers such as the African dwarf are much more difficult to get into binocular vision. A number of species have wide sub-Saharan ranges and occupy suitable watery habitats within many vegetation zones, be it tropical forest or savanna. Others occupy streams and rivers only within the tropical forest belt and the common kingfisher is restricted as a breeding resident to the extreme north-west of the continent. During the northern winter these residents are joined by substantial numbers of non-breeding migrants. They concentrate along the coast and favour quiet waters such as in estuaries and mangrove swamps.

In the case of the so-called woodland, or dryland, kingfishers, some have a strictly tropical range, such as the chocolate-backed, but others, for example the grey-headed and woodland kingfishers, are much more cosmopolitan in their habitat tastes. Kingfishers with only a marginal presence in Africa include the beautiful white-breasted, whose population is limited to a handful of birds on the upper Red Sea and near Cairo. White-collared kingfishers have a toehold in the mangroves and muddy shores associated with them along the Red Sea coast; elsewhere they occur eastwards across the Gulf of Aden to several island groups in the Pacific.

As far as diet is concerned the dryland kingfishers take a wide range of invertebrates and also lizards, frogs, small snakes, tiny birds and mice. Those occurring in areas where driver ants are present follow the columns and snatch up small animals fleeing these rapacious hordes. White-collared kingfishers catch insects, fish and especially crabs. We have spent some time watching these birds hunting over mangrove mud-flats where they were mainly concentrating on the abundant fiddler and *Sesarmid* crabs.

Most species carry the prey back to the hunting perch and proceed to give it a good pounding before juggling it into the right position for swallowing. Some species, for example the pied kingfisher, hover above the water with rapid wing-beats and with the bill held vertically to the water. Once the prey is spotted the bird executes a neat and rapid plunge. In some areas pied kingfishers reach fairly high densities and it is quite a sight when a dozen or so are busy hunting. This is also one of the few kingfishers that roosts communally, with more than 200 in a group having been recorded. Most kingfishers, however, are territorial and do not tolerate intruders.

All kingfishers are hole-nesters, usually in a burrow that they have excavated themselves. One of the more unusual nest sites is that of the chocolate-backed kingfisher. This species, as well as the blue-breasted kingfisher, excavate their nest-burrows in the tree colonies of termites and arboreal ants. Many others excavate their burrows and nest cavities into earth banks, and on occasion the large mounds of terrestrial termites. The nest cavities of kingfishers are never deliberately lined but soon become cluttered with small bones, scales and other inedible food remnants, mainly from regurgitated pellets. The woodland kingfisher, and others, will take over holes in trees chiselled out by woodpeckers and barbets, as well as natural tree holes. The five to eight, white eggs – typical of hole-nesting birds – are incubated by both parents, and both feed the chicks. ■ SEE PAGE 160.

BIRDS IN THE RAINFOREST

As the vegetation of a rainforest is stratified, so is its birdlife! Many forest birds are relatively small, low wing surface area being an advantage in dense vegetation. There are only two levels in the forest where large wings are not a hindrance: on the forest floor and in the tree-tops. It is in the forest canopy where such species as hornbills, parrots, trogons and raptors concentrate to feed and to fly without hindrance over the trees. On the forest floor several species of guineafowl and fran-colin forage and breed. Most birds in the African forests travel singly, in pairs or in mixed-species feeding parties – referred to as "avian armies".

Because grain is a limited food resource in the forest, seed-eaters largely give way to small fruit- and insect-eating birds. Although several species of weaver occupy parts of the forests, unlike their woodland and savanna relatives they feed on insects and a variety of wild fruits. True insect-eaters such as flycatchers, bulbuls and robins are the typical forest dwellers, dominating most small bird flocks. The mixed-feeding flocks, also known as bird parties, may be made up of several dozen individuals of several species. They move slowly through the forest with each species hunting its favoured prey. Some are ground foragers, others hop through the undergrowth, still others feed at higher levels, and then there are the aerial hunters. There is one major advantage for these large mixed-bird flocks: they create a considerable disturbance among the vegetation. Many insects that would normally remain hidden when a single bird is foraging are alarmed enough to leave their hiding places when these large flocks "blitz" their micro-habitat. A strategy used with considerable success by these mixed-bird flocks is to follow driver ant columns. These voracious invertebrate predators devour insects too slow to flee their columns, those agile enough to escape the ants are taken by the birds.

The wood nuthatch has only a "toehold" in extreme northwestern Africa.

A brown-hooded kingfisher.

Nuthatches

FAMILY SITTIDAE

Only two nuthatches occur in Africa, both in limited areas of the north-west. The Eurasian wood nuthatch, elsewhere widespread, is found in a small section of the Atlas Mountains in Morocco, and the Kabylian nuthatch, sometimes called the Algerian nuthatch, is a recently discovered African endemic. It is only known to occur in the coniferous forest on the Petite Kabylie and, amazingly, it was overlooked by Palaearctic ornithologists until 1975.

These small, short-tailed, stocky birds are the only tree- or rock-dwelling birds that frequently descend head down. The bill is fairly long and sharp-pointed, the feet are strong and the toes are equipped with long claws for clinging to steeply angled and vertical surfaces. Nuthatches feed on a variety of invertebrates as well as plant material, mainly seeds and nuts, which may be wedged into cracks in tree bark and then hammered open with the bill. ■ SEE PAGE 162.

The trogons, in this case a Narina, are everybody's idea of how a tropical bird should look. (Photo: JJ Brooks)

Trogons

FAMILY TROGONIDAE

These magnificent birds epitomise "everybody's idea of a tropical bird". Trogons have their highest diversity in the Americas and Southeast Asia, with just three out of perhaps as many as 38 species occurring on the African continent. These brightly coloured forest birds all nest in tree-holes.

Apart from their showy plumage colours, trogons have small and weak legs, and so-called "yoke-toes". Although several other bird groupings have zygodactyl feet (two toes forward and two back), trogons are unique in that it is the inner, or second toe, that is to the rear, and the two front toes are usually joined at the base. A strange feature of these magnificent birds is their almost unbelievably loosely attached feathers, and skin so soft and fragile it has been likened to wet tissue-paper. All are long birds, made more so by the substantial tail, but all African species are under 100 g. The wings are short and rounded, heads large with well-developed eyes, and females are duller than their showy male counterparts. Despite their bright and boldly marked plumage they are difficult to observe in their dark forest habitats. Their calls are often the only indication of their presence; the vocalisations are rather limited but distinctive, of the hooting-cooing type. The short but broad bills, with

bristles at their bases, are an adaptation for their mainly insectivorous diet. When foraging they usually concentrate on the middle canopy of the forest and adopt a wait and pounce strategy, taking their prey from the vegetation in flight. At least in the Narina trogon some small vertebrates such as chameleons and geckos are eaten.

Usually single birds or pairs are encountered and at least in the breeding season they are territorial but this may extend to other times as well. Trogons are largely sedentary with no long-distance movements, but there may be some local seasonal and altitudinal movements. As is typical of hole-nesting birds, the eggs are white, usually numbering two or three per clutch. Both parents take on the task of incubation and chick raising.

Although the widespread Narina trogon has a fairly continuous distribution in the tropics, the bar-tailed trogon has a similar range but is extremely patchy. This can in part be explained by its favouring montane forests, but it may well be more widespread than present distribution records indicate. The bare-cheeked trogon is found in two apparently isolated populations in the upper Congolean forest belt but it is highly likely that they are linked. ■ SEE PAGE 160.

Wood-hoopoes

FAMILY PHOENICULIDAE

The eight species, in two genera, of wood-hoopoe are all immediately recognisable as belonging to this purely African family. All are small to medium-sized birds distinguished by their long, pointed tails, plumage of iridescent greens, blues and purples with white in some species and long, slender down-curved or straight bills. The feet are fairly small but each toe has a long and sharply curved claw, an adaptation to their arboreal life style.

Although two species are mainly solitary or occur in pairs, the others are usually seen in small parties from four to 12 in number. Their food is gleaned from tree trunks, branches and vegetation tangles, as they creep and hop around in a somewhat rodent-like way. Their agility allows them to explore even the underside of horizontal branches with ease, the tips of their bills constantly probing under bark, leaves and in rotten wood. The tail, like those of woodpeckers, is used as a support on vertical logs. Small insects and other invertebrates make up the bulk of their food but small quantities of berries and seeds are taken from time to time.

The more social wood-hoopoes are noisy and make their presence known but the rest are much quieter and secretive. Although the behaviour of some species is reasonably well known, little is known about the forest, black-billed and violet wood-hoopoes. A few species occupy forest habitats, others wooded savanna mosaics, the remainder preferring fairly dry acacia woodland. Only the green wood-hoopoe, with a wide sub-Saharan distribution, has been studied in any detail. It forages in noisy groups and each party vigorously defends its territory throughout the year. This defence includes impressive displays and much calling. At night they roost in tree holes, males and females separately most times. They have an interesting social system, in that should a member of the group be taken by a predator or disappear for some other reason, birds of the same sex emigrate to the group from flocks occupying adjacent territories. Breeding pairs are assisted by "helpers", from one to as many as 10 non-breeding birds, which are usually from previous broods of the pair. As the breeding season approaches the breeding pair spend much time together grooming or quietly perched. Copulation is frequent and for a bird very long at up to two and half minutes per session. Once the clutch (from two to five in this species) has hatched, the male and the helpers bring food to the breeding female and she then feeds it to the young. Flock members continue to feed the young birds for several weeks after they leave the nest, and this brood will act as helpers with the following batch of young. For up to five years these birds may continue to be helpers to their parents. In the unlined nest cavities, nearly always in natural tree holes or old woodpecker or barbet nests, the youngsters dwell in the same smelly and rather repulsive conditions as the hoopoe. ■ SEE PAGE 160.

Above: *A scimitar-billed wood-hoopoe emerging from its nest-hole.*
Right: *The green, or red-billed, wood-hoopoe occurs in small but noisy flocks.*

Above: *A bearded woodpecker at the entrance to its nest.*
(Photo: Clem Haagner, ABPL)
Right: *The ground woodpecker is endemic to South Africa.*
(Photo: Warwick Tarboton, ABPL)

Woodpeckers, piculet and wrynecks

FAMILY PICIDAE

Africa can boast a tally of 32 woodpeckers, one piculet – a mini-woodpecker – and two wrynecks, the majority being sedentary endemics. With one outstanding example, woodpeckers have a tree trunk and branch "fixation", spending most of their lives aloft foraging, roosting and breeding in or on them. The one that is different is the fairly large ground woodpecker, a near South African endemic. This bird has adopted a terrestrial life style, doing nearly all of its foraging on the ground and excavating its nest-hole in earth banks and in old farm buildings.

Woodpeckers are the true masters of tree trunks, having no equals in the realm of forest and woodland. They are superbly adapted to their arboreal life style, with a stiff tail that is used as a prop on vertical surfaces, long and strong toes armed with sharp, curved nails, and a straight and sharp or chisel-pointed bill mounted on a thickened skull that helps absorb the shocks from tapping. Their tongues are very long and extensile, with backward-pointing hooks at the tip. On opening the tunnel of a grub the tongue snakes in to spear the hapless meal. It is believed that the saliva of some species contains a sticky substance, a further aid to prey capture.

Woodpeckers occur across much of Africa. Many species are endemic and a few are very localised. In this category we find the Knysna woodpecker, which occurs only in southernmost South Africa, and Stierling's woodpecker, a rare inhabitant of *Brachystegia* woodland centring on southern Malawi. Three species that have wide Eurasian ranges also have a limited toe-hold in extreme North Africa, namely the lesser- and great-spotted woodpeckers, and the green woodpecker. One of the most widespread is the small but pugnacious cardinal woodpecker. We got to know a pair very well at one of our previous homes, not always in the way we would have liked ... The male took a fancy to our large, old sash windows, frequently descending to tap out the putty and leave holes and gaps in the woodwork. Even when we were sitting on the veranda, down he would come and on most occasions ignore our protests.

Woodpeckers are handsomely marked birds, most having variable amounts of red, white and black on the head, green or brown backs with abundant speckling, and barring and streaking fore and aft. The olive woodpecker lacks spotting and its plumage is more uniformly coloured than in the other species. Their coloration and markings make them difficult to spot in their tree habitat, and in addition many species keep trunk or branch between themselves and the human observer. Their flight is rather heavy and undulating, with a few wing flaps carrying them upwards, followed by a downwards swoop.

Many of Africa's woodpeckers have received scant attention from ornithologists because they are seldom seen, in particular those that inhabit dense lowland rainforests. Although tree foraging is prevalent in most woodpeckers, a few species, such as the fine-spotted, do on occasion hunt for their insect prey on the ground. The green woodpecker spends quite a lot of its foraging time on the ground, particularly around ant colonies. All species feed mainly on insects, in particular ants and their larvae; other invertebrates such as spiders are taken on occasion. Although this aspect is poorly known, at least some woodpeckers are omnivorous and take some plant food. A few woodpeckers are known to join mixed-species foraging parties.

At least in the few known species, most woodpeckers live as monogamous pairs within a territory, throughout the year and not just during breeding. Nesting and roosting holes are excavated in dead wood, either in the trunk or branches, but the buff-spotted woodpecker makes holes in the nests of carton ants, which are attached to branches. The clutch of white eggs numbers from two to eight but many woodpeckers lay only two or three eggs.

The two wrynecks take their name from the manner in which they turn their necks while foraging for their predominantly insect food. The prey consists mainly of ants and their larvae, which they take from trees and commonly from the ground. The plumage, a mottling of grey, black, white and brown, is very well camouflaged and makes spotting these birds difficult. Unlike woodpeckers, wrynecks have soft tail feathers and therefore cannot use the tail as a brace when moving on vertical tree trunks. They also differ from woodpeckers in that they sit crosswise on a branch, as opposed to in line with it. The pointed, slightly down-curved bill is used for probing and not for wood-chiselling. Unable to excavate their own nest-holes the wrynecks make use of natural cavities, or old woodpecker or barbet holes.

The rufous-necked wryneck, an African endemic, has a wide but strangely patchy distribution and as it tends to be nomadic out of the breeding season it has perhaps been overlooked in some areas. Eurasian wrynecks have a wide Palaearctic range, including a breeding population in north-western Africa, but non-breeding northern migrants occur in a broad swathe from the Atlantic almost to the Red Sea and to south of the Sahara. The Eurasian wryneck usually lays from five to 10 eggs but the rufous-necked between one and five eggs.

Africa's only piculet looks and generally acts like a woodpecker, which it is in diminutive form, but like the wrynecks it sits crosswise on branches and the soft tail feathers are not used as a prop on vertical surfaces. Its range is largely restricted to the Congolean forest block, where it can be solitary, in pairs or small parties. It forages mainly in the undergrowth for its insect prey. Insects may be picked off the vegetation, but the bird also splits stems to obtain grubs, as well as chiselling at bark like its larger cousins. ■ SEE PAGE 161.

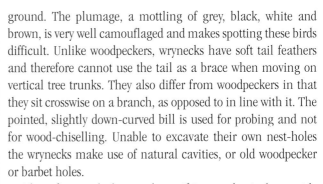

Left: *The golden-tailed woodpecker.* (Photo: Hein von Horsten, ABPL)
Right: *A female Bennett's woodpecker* (Photo: Brendan Ryan, ABPL)
Below right: *The ground woodpecker, a southern African endemic, has forsaken the trees to forage on the ground and nest in burrows in banks.*
Below left: *The rufous-necked wryneck feeds mainly on ants and their larvae.* (Photo: John Carlyon)

10

LBJs

"Little brown jobs" is birding parlance for those generally small, nondescript and difficult to identify birds that may send a shiver down the spine of even the most hardened birdwatcher or twitcher. Many LBJs are in fact shades of grey, green or yellow, but all appear dull at a distance. They are found in all habitats that Africa has to offer, from forest to desert, coastal plain to high mountains. The vast majority have received little attention from ornithologists, except maybe to fill museum drawers and cupboards. It is almost certain that those new species awaiting discovery in the African tropics will primarily fall within this non-scientific division. Shall we forget the frustrations of those blurs of brown fleetingly observed in some of the marshes of western Tanzania, or the forests of Uganda? To remain forever unidentified! Fortunately, many species have distinctive calls that are useful aids to identification, and some exhibit characteristic behaviour. Among the ranks of the LBJs are some fine songsters. Indeed, if human, they would attract the tag of amateur musicians; others utter harsh, grating calls – positively unmusical. Many are African endemics but a substantial number enter the continent as breeding and non-breeding migrants. There are those with very restricted distributional ranges; others occupy extensive areas that fulfil their specific habitat requirements.

Accentors

FAMILY PRUNELLIDAE

The hedge accentor is a vagrant to the north and the Alpine accentor has limited breeding populations in the Atlas of Morocco and also occurs as a scarce winter migrant over a slightly larger area. These small terrestrial birds are rather sparrow-like in appearance but they have the more slender, pointed bills of insect-eaters. ■ SEE PAGE 163.

Left: *The red-eyed bulbul is frequently seen in southern Africa.* (Photo: John Carlyon) **Right:** *The yellow-vented bulbul, a mainly Arabian bird, enters the Sinai and possibly the Horn of Africa.*
Above: *Melodious lark.* (Photo: Rob Ponte, ABPL)

Bulbuls and greenbuls

FAMILY PYCNONOTIDAE

With a few exceptions this is a dreaded group for professional and amateur birders alike. Africa is home to 67 species, half of the world's total. A few species are well known and common in gardens, parks and even suburban walkways, but the vast majority are secretive, skulking and little known.

Bulbuls and greenbuls have a patch of hair-like, vaneless feathers located on the nape, but this is not a field feature. Most are thrush-sized and smaller. They have fairly short and rounded wings, and moderately long tails that may be squared, rounded or in a few species slightly forked. The bills in most species are fairly slender and slightly down-curved. Several species, in particular the bulbuls, have crests on the top of the head that normally lie flat but are raised when the bird is displaying or alarmed. With regard to plumage, the vast majority are rather sombrely clad in browns, greens and greys, with a few exhibiting patches of yellow, red or white on the head or in the vent region. In general the body plumage is fluffy and soft, especially on the lower back. Males and females look alike but males are slightly larger, a feature that one would not pick up in the field.

Although a few species have been studied in detail, many members of this family are little known. The greatest diversity of species occurs in the tropics, where most are associated with forest and woodland habitats. Several, particularly in the genus *Pycnonotus*, are commonly associated with humans and their gardens. Most species are arboreal but several also forage on the ground, where they seek out their mixed diet of mainly fruit and insects. In general they are territorial when breeding and some live in groups that jointly defend a home range. Others move around in small, noisy bands that are often inquisitive.

As is to be expected for such a large group of birds, some species have extensive African ranges but many have more restricted distributions. In this latter category fall such species as the Cameroon greenbul, Toro olive-greenbul and Sassi's greenbul. Most African species are endemic and sedentary; some show nomadic tendencies outside the breeding season which are probably related to food availability and abundance. The nests are of two basic types: most species construct "typical" bird nests which are cup-shaped and placed in a tree fork; a few species suspend hammock-style nests from a branch or twig. Most bulbuls in the tropics have clutches of just two or three eggs but they often rear more than one brood in a season. Up to five eggs (usually fewer) may be laid by species in more temperate zones. ■ SEE PAGE 162.

Above left: *Sombre greenbul sitting on eggs.* (Photo: John Carlyon)
Above right: *Yellow-bellied greenbul show a strong preference for coastal and riverine forest and bush.*
Far right: *The yellow-vented bulbul is common within its very limited African range.*
Right: *Black-eyed bulbul.*

THE PASSERINES: BIRDS THAT CAN PERCH

This staggeringly vast order, Passeriformes, incorporates a bewildering array of species which account for approximately 60% of all living birds. How many families should be recognised is a matter of considerable debate, as well as how they relate to each other.

The name of this order is a reference to their feet, which are adapted for perching. Commonly called passerines, the perching birds have four unwebbed toes, three facing forward and one backward, with the hind toe often being the strongest. They also differ from birds in other orders in the structure of their palate bones, a reduction in the number of the cervical vertebrae and distinctive spermatozoa, among other characteristics. The young of nearly all species are altricial, meaning they hatch blind and naked and are reared to fledging in the nest, therefore being totally dependent on the parents. However, in Africa two groups are brood parasites: the whydahs and widowfinches.

Passerines occur on every landmass except Antarctica. In Africa they range in size from tiny, for example sunbirds and prinias, to the medium-sized members of the crow family.

By far the largest suborder, the Passeri, are classed as "true oscines" or "true song birds". They have the greatest number of syrinx, or voice-box, muscles and are the "classical musicians" of the avian world; not for them the discordant cacophony of the "moderns" (such as red-billed wood-hoopoes) or the "heavy rock" of hadedas, trumpeter hornbills *et al*! Nevertheless many songbirds do not live up to their title, but a greater number do, as anyone who has been roused by the musicality of Heuglin's robin-chat or the perfect duet of bokmakieries will testify.

It is the music of forest songbirds, which rarely show themselves to the human observer, that helps us (hopefully) to identify them, but the raucous variety of sound at times overwhelms. In a glade in the Budongo forest of western Uganda we sat listening in awe to the avian choir and it brought to mind a segment of poetry by John Logan, in *To the Cuckoo*:

Thou hast no sorrow in thy song,
No winter in thy year.

Far left: *Red-winged bush lark.* (Photo: Clem Haagner, ABPL)
Left: *Like many lark species the rufous-naped has a bewildering array of 23 subspecies. This lark frequently sings from a prominent vantage point.*

Larks

FAMILY ALAUDIDAE

The majority of lark species occur in Africa. These primarily terrestrial songbirds have cryptically coloured plumage which blends in with the substrate on which they live. The latest African tally, according to some, is 73 species, a number of which have been described only in the last 20 years.

One anatomical feature that sets larks apart from all other passerines is that the back of the tarsus is rounded and scaled, instead of sharp and unsegmented as in all other families. Several internal anatomical features also are restricted to this family. Many species are known for their impressive songs or display flights. In most species the wings are proportionally long and pointed. The tail is short to moderate in length and the bill may be long, slender and pointed to short and heavy, and just about everything in between. Many have crests, such as the crested lark (what else?), the skylark and thekla lark, and the horned shore lark has two distinctive "ear-tufts". Most species walk or run on the ground, as opposed to hopping.

These are birds of open and lightly bushed savannas, desert and semi-desert; they avoid dense cover or forest. Several species have benefited from the destructive activities of people: the clearing of land for agriculture, and overgrazing and trampling by their herds. While this has created suitable habitat for some species, it has had a negative impact on certain localised species, such as Botha's lark in South Africa, a denizen of upland grassland. A number of species have extensive distributional ranges, including the flappet, hoopoe, bar-tailed and red-capped lark and the chestnut-backed finchlark, but many more have limited to very restricted ranges. Some lark species are known from just a small number of specimens.

Most larks construct cup-shaped nests on the ground; depending on the species the nest is positioned at the base of grass tuft or bush, close to a rock or completely in the open.

Larks of the genus *Mirafra* use plant materials at the sides and back of the nest to create a dome. Nests are usually well lined. Clutch sizes range from two to six eggs, and it is common for some of those species studied to have two broods in a season.

Many larks are great songsters and they combine this with impressive courtship flights. The best known has to be the Eurasian skylark, which has moved poets such as Wordsworth, Shelley and Tennyson; even Blake in *Auguries of Innocence*:

> *A skylark wounded in the wing,*
> *A cherubim does cease to sing.*

The Eurasian skylark has one of the lark world's greatest distributions but in Africa it is restricted to the far north. This "blithe spirit" of Shelley's verse with its spiralling courtship flight and bubbling song is the supremo of lark performers. Species such as the flappet, clapper and collared larks include wing-clapping in their aerial displays, and a few incorporate singing from a perch into their repertoire. Many males sing in flight, with the spectacularly marked hoopoe lark putting on a fine display. He flies up from a bush or tuft with spread tail and fluttering wings, and at a few metres above the ground he does a roll and goes into a seeming death-plunge back to the perch. The displays and singing go on intensively in many species throughout the breeding season, and some even continue to call into the night. A few days after rain has fallen, when the vegetation is putting on a new face, lark activity is usually intense, with several species competing for "air-time". We can recall driving through arid areas on many such occasions, delighted by the "lark extravaganza".

Although most species pair off and are territorial when nesting, out of the breeding season many larks form single-species, or even mixed-species, flocks. Some are sedentary or locally nomadic, others undertake migrations to and from the breeding grounds. The mainly plant-eating thick-billed lark of North Africa forms into large numbers of small flocks.

Larks are by and large omnivorous, taking both insect and plant food, the latter being mainly seeds, but the nestlings are usually given a diet dominated by insects and other small invertebrates. ■ SEE PAGE 161.

THREATS TO MIGRANTS

Many different species of bird enter and leave Africa as seasonal migrants, primarily from the Palaearctic Region but also from the Oriental Region. The three principal migrant flyways into and out of Africa are the western route across the Strait of Gibraltar, the central route across the upper reaches of the Red Sea, and the south-eastern route across the Strait of Bab-el-Mandeb in the Red Sea, the narrowest land link between Asia and eastern Africa.

A noticeable decline has been recorded over recent years in the population numbers of several species that make these great journeys. Relatively few land birds traverse the entire length of the African continent, although there are exceptions, such as the barn swallow (*Hirundo rustica*) and white stork (*Ciconia ciconia*). A great number of migrant species choose to overwinter in the Sahel belt which separates the harsh, bird-unfriendly Sahara Desert from the equatorial zone to the south. Still others spend at least part of their time in the Sahel both on their southwards and northwards journeys, replen-ishing their reserves. The bulk of the redstart (*Phoenicurus phoenicurus*), Eurasian wheatear (*Oenanthe oenanthe*) and whitethroat (*Sylvia communis*) populations remain within the Sahel belt for the northern winter. This harsh wintering ground may appear to be a strange choice, but when the birds first arrive the Sahel is usually lush and green from the rains. Even when the dry season arrives and the vegetation dries out, many species can still glean a living here. Others use it purely as a staging post and move southwards.

Those birds that overwinter in the Sahel seem able to exploit feeding situations that are not utilised by African species. A different case is that of the migratory Eurasian wheatear: it exploits the same feeding niche as the local Heuglin's wheatear (*Oenanthe heuglini*) but competition is prevented by the African species migrating southwards just prior to the arrival of its Eurasian cousin. It returns to the Sahel only once the northern species has left again for its breeding grounds.

Declines in the populations of a number of northern migrants have been attributed to adverse conditions in their African wintering grounds, particularly the Sahel, which is a fragile ecosystem. Long periods of drought in the 1970s and 1980s resulted in devastating overgrazing and massive cutting of vegetation to provide food for tens of thousands of head of domestic stock, as well as fuelwood. This was aggravated by previously nomadic peoples now concentrating around foreign aid centres and the few wells. These and other factors so damaged this region that it is unlikely, given current conditions, to recover fully. Although redstarts and whitethroats have been badly affected by these changes, others such as nightingales and garden warblers do not seem to have suffered any major population declines. This can be explained, in part, by the fact that they are itinerant visitors, only remaining in the Sahel for a short period. Arriving in the Sahel early in September, by the beginning of November the nightingales and garden warblers have already moved southwards to the fringes of the higher-rainfall forests of West Africa.

Two other northern migrants, the spotted flycatcher and willow warbler, arrive just to the south of the Sahara Desert earlier than most other migrants when it is still very green, but by the middle of October they continue their journey to catch the rains to the south of the equator. The flycatchers, in particular, will linger around productive feeding grounds and may only arrive in their primary wintering grounds in November, well over three months after they left their summer breeding grounds.

Before the long return flight to the northern breeding grounds both the species overwintering in the Sahel and the itinerants such as the spotted flycatcher have to build up reserves for making the perilous desert crossing. By this time the northern tropics are some six months into the dry season and even though the Sahel is now parched, it is yet again a crucial "refuelling" stop-over because there is an abundance of fruit on the bushes and trees – at least there used to be! The numbers of fruiting bushes have been decimated, as have insect populations reliant on the vegetation.

Pipits, longclaws and wagtails

FAMILY MOTACILLIDAE

This group includes some of the most endearing and some of the most frustrating of birds. Many of the brown and mottled pipits are at the top of the scale of hard to identify LBJs. The longclaws have useful identification patches on their throats, although they also have the frustrating habit of keeping their nondescript backs to the observer! Some of the wagtails are easy to identify but there are those, such as the grey and yellow wagtails, that have hordes of often difficult to separate subspecies.

Seven wagtails, 27 pipits and seven longclaws grace our waterways, grassland and lightly bushed country. Many species are endemic, others undertake long-distance migrations, some are non-breeding visitors. In general the smallest are the wagtails and pipits, with the longclaws being a little larger. The bills are of moderate length and pointed. Many of these birds have quite long legs and well-developed toes and claws, those of the longclaws being the most developed. Their long toes and claws are an adaptation to their terrestrial way of life. It is of interest that the longclaws and the unrelated meadowlarks — very similar in appearance — of North America offer an excellent example of parallel evolution. The habit of tail wagging is not restricted to the wagtails but is also found in several of the pipits, although usually less pronounced.

Many species exhibit the so-called song flight display, which is also a feature of the larks, as described above. The pipits take their common name from the nature of their calls. Small invertebrates make up the bulk of the diet of all species, most taken on the ground while the bird is walking or running. Watching a wagtail feeding can be a dizzying exercise, as they meander this way and that, pecking at the ground here and then suddenly dashing off to snatch a flying insect there.

Passerines do not have an extensive fossil record but it is believed that the ancestry of this family goes back at least to the upper Oligocene, some 30 million years before present. A new species discovered in South Africa in 1995, the long-tailed pipit, had obviously been around for a long time but was overlooked. Many wagtails and particularly the pipits have seemingly disjointed ranges in Africa but we feel the reason is that these birds are being overlooked and under-recorded in many cases. A few species are highly localised, including the Sokoke pipit which is restricted to a few forest pockets in coastal Kenya. It has been estimated that there are far less than 10 000 pairs. With the exception of the yellow-throated longclaw, other members of the *Macronyx* genus also have restricted and localised ranges.

In the breeding season birds are usually seen in pairs but at other times many species form flocks, particularly for migration and at the wintering grounds. Most wagtails roost off the ground, and migrating birds often congregate in their thousands at roost sites. In the case of the yellow wagtail numbers at primary roosts may number in the tens of thousands. Although migration patterns are long established, every now and then a bird throws a spanner in the flight path works. The citrine wagtail is a vagrant to a few sites along the African Red

Far left: *Yellow-throated longclaw.*
Above: *Grassveld pipit.*
(Photo: Richard du Toit)
Left: *Although a few races of the yellow wagtail breed in North Africa, still others enter the continent as non-breeding migrants.*

Sea coastline but in 1998 a single individual arrived at the southern coast of South Africa and proceeded to make itself entirely at home, even courting the local Cape wagtails.

Wagtail nests are open but often fairly deep cups of plant material placed on the ground or among vegetation tangles, in the case of the more water-associated wagtails. Clutch size is variable but usually between two and seven eggs.

Unlike many of the pipits and longclaws, several of the wagtail species become very trusting of humans. For example, the Cape wagtail is common in our home village where it associates strongly with gardens. One can be watering the plants or digging over one of the beds and there they are, competing with Karoo robins for insect snacks. African pied wagtails likewise show little concern for people. ■ SEE PAGE 165.

The spotted prinia is common over much of its rather limited range. (Photo: John Carlyon)

Warblers

FAMILY SYLVIIDAE

This is a very large family, communally referred to as Old World warblers, many members of which cause the average birder to go into a cold sweat and turn to the bottle for inspiration. At last count 228 species were recognised for the African continent, the vast majority being endemic residents, although there are a number of intra-African and Palaearctic migrants. This large family is divided into separate families by some authorities but the arguments are complex and we have plumped for leaving them all under the same umbrella. Perhaps laboratory work in the future will elucidate matters but we have a feeling that instead it will probably complicate the issues even further. As it is, 49 different genera are recognised, and although a few are distinct and easy to identify in

the field, many are difficult in the extreme; even the experts have their hands full. Apart from those species carrying the common name of warbler, this vast assemblage also includes the whitethroats, crombecs, hyliotas, camaropteras, prinias, cisticolas and the 26 apalises.

The vast majority of species are tiny to small and dull-coloured, and members of certain genera are very similar in overall appearance. They are usually highly active little birds – as many a bird-photographer will attest – flitting among the undergrowth or in the canopy of forest and woodland instead of holding still in a suitable pose for their picture. Many species also inhabit thickets and tangles, reed-beds and marshes, expanses of dry and moist grasslands and even semi-desert where there is some vegetation cover. In general, they have rather thin and weak legs and feet, and thin and fine, sharp-pointed bills. The tail may be anything from short to long; the wings are usually fairly short and rounded.

Several species are fine musicians, but many have their calls explained by their very names: clamorous reed warbler and rattling, croaking, wailing, chattering, bubbling and zitting cisticolas. At least their distinctive calls, frequently uttered during the breeding season, serve as a guide to the identity of these lookalike and often invisible birds.

The vast majority are insect-eaters but a few include vegetable matter in their diet, in particular small berries. In fact as I am writing this I am watching a spotted prinia picking aphids from the stems of a low shrub just 5 m from the window. Although most warblers pick their insect prey from vegetation, several also opportunistically hawk for these invertebrates, some species being better adapted to this than others. Few species are recorded as taking small vertebrates but larger warblers, such as the Cape reed warbler, are known to catch frogs. The yellow flycatcher-warbler is a keen caterpillar predator;

another ardent moth and butterfly larvae specialist is the bar-throated apalis. Apart from eating small fruits, a few warblers have been noted taking flower buds and even sipping nectar in the manner of sunbirds. Although the cisticolas seem to be more exclusively insectivorous than other groups, there are of course exceptions. The rattling cisticola has been recorded as taking nectar from aloes, as has the tawny-flanked prinia. In a surprising number of species we really do not know the diet but dare to presume!

The nests are usually small cups, or in many cases domed structures, often intricately woven from plant fibres, wool, hair or spider webs. These are designed by fine avian architects but the title of master builder should go to the tailorbirds. These are members of a predominantly Asian group of less than a dozen species, but two, the long-billed and African tailorbirds, pose something of a mystery to zoogeographic experts. Rather plain, nondescript birds — in other words the typical warbler — they have the distinction of making impressive nests. The nest is constructed within a cup by sewing the edges of one or two leaves together. ■ SEE PAGE 164.

Right: *The long-billed crombec has a characteristic "hanging bag" nest which is usually bound with spider web.* (Photo: John Carlyon)
Below: *African marsh warbler.* (Photo: Brendan Ryan, ABPL)
Far right: *Cape reed warbler.* (Photo: Brendan Ryan, ABPL)

Wrens

FAMILY TROGLODYTIDAE

A species on the fringe of its vast Palaearctic range is the wren, affectionately known as Jenny Wren, or King of the Hedges — a grand name for a tiny but pugnacious, short-cocky-tailed, cryptically coloured bird that builds a large domed nest and thereby acquired its much less grandiose scientific name, meaning "cave dweller". In Africa this little bird occurs only around the Mediterranean seaboard.

According to folklore, all of the birds gathered to decide who should be the ruler of the avian kingdom. It was decided, in good democratic fashion, that the bird who flew the highest would be declared king. All the birds ascended towards the heavens and just as the eagle was about to proclaim his triumph, a hitch-hiking wren flew a little higher from his back, calling out his newfound majesty. ■ SEE PAGE 163.

11

Insect-eaters

This chapter covers species whose diet consists primarily of insects and other invertebrates, although many also eat some plant food from time to time. Of course invertebrates feature at different levels in the diets of many other birds. A few birds, such as the drongos and flycatchers, could also fit into the chapter on aerial feeders, but we have chosen to include them here. Here we find Africa's finest songsters in the robin-chats and thrushes, some of its colourful representatives in the orioles and some of its noisiest in the babblers. Most members of the crow family are omnivores, and not rarely active predators, but such tidbits as locusts, termites and beetles are taken with relish. Here we also encounter those miniraptors, the shrikes and kin, many of which are pugnacious and active. Others are skulking but compensate for their rare appearances with fine song. Although most birds eat their food on the spot, some of the shrikes impale what is surplus to their needs on thorns or barbed wire. The starlings range from solitary species to great seasonal flocks of wattled starlings. The birds in this chapter are often referred to as songbirds; they are passerines, or birds capable of perching.

Babblers

FAMILY TIMALIIDAE

This is an amazingly diverse passerine family, with species ranging from tiny to the size of small crows. A total of 36 species occur on the African continent, in all incorporating 11 genera, out of a world total of approximately 250 species. With the exception of one species in North America, members of the babbler family are restricted to the Old World. It is often felt that this group should be divided into several different families, taking into account that at present six tribes are recognised, of which three have representatives in Africa and a fourth is represented in Madagascar.

Variable in size and overall appearance, members of this family are united in certain features. These include fairly long and strong legs and toes, and short and rounded wings. Most have rather dull but fluffy plumage although there are a few exceptions. Most species are social and noisy, both at perch and when foraging; most are arboreal but some are terrestrial in their foraging behaviour. Nearly all occur in woodland or forest, although those in drier areas may occupy sparsely wooded expanses, usually with a scattering of copses and small thickets. It is their noisiness and constant chattering that give many

Above: *The buff-streaked chat is a southern African endemic, mainly associated with rocky hill slopes.* (Photo: John Carlyon)
Left: *The chinspot batis is one of 19 batis species in Africa.* (Photo: Richard du Toit)

Arrow-marked babbler. (Photo: John Carlyon)

of the species their common names. Much of their food is made up of insects and other small invertebrates, but in many species their diet is supplemented by small fruits and berries.

In general they are solitary nesters, but in many cases the parent birds are assisted by one or more helpers. Depending on the species and situation, helpers may assist in nest-building as well as feeding nestlings, and this is best developed in babblers of the genus *Turdoides*. Nests may be open cups or domed structures with side entrances.

Although a few species are quiet and unobtrusive, many, in particular the genus *Turdoides*, are noisy. They move about in small groups that seem to be almost continuously communicating with each other. During foraging bouts on the ground, throwing over leaf-litter, or in the undergrowth, they will suddenly start calling in chorus, then quietly continue to pick and peck. Many species when calling and displaying, spread and depress the tail feathers and expose the rump with fanned wings held downwards and quivering. ■ SEE PAGE 162.

Broadbills

FAMILY EURYLAIMIDAE

Four broadbill species occur in Africa. Only the African broadbill has a fairly wide but disjointed distribution; the most localised is the African green broadbill, centred on the Itombwe range and the mountains west of Lake Kivu.

Broadbills have a simple syrinx, partial joining of the front toes and 15 instead of 14 neck vertebrae. Because of these characteristics and their generally scattered distribution it is often held that they are an ancient group that could well be in evolutionary decline.

They are smallish, plump birds with disproportionally large heads set on short necks and, as their common name implies, their bills are broad, stout and hooked at the tip. These forest dwellers feed primarily on insects and other small inverte-

brates, and at least the rare African green broadbill eats some plant food as well. Nothing is known of the diet of the grey-headed broadbill. Prey is caught mainly by plucking it from the foliage but hawking for flying insects has also been recorded.

The nests of broadbills are beautifully woven, hanging purses that may be held to a branch or twig by a woven "thread" of the same plant material. The rufous-sided and African broadbills attach their nesting purse directly to a vine or twig without the suspending string but material may hang below the nest, looking like a straggly beard. Fungi, lichens, moss and leaves are usually incorporated into the nest, making it difficult to find. A surprisingly small clutch of one to three eggs has been recorded in the very few nests ever found. ■ SEE PAGE 161.

Creepers

FAMILIES CERTHIIDAE, TICHODROMADIDAE AND SALPORNITHIDAE

Most authorities place the two tree creepers, one wall creeper and the spotted creeper respectively in the three separate families given above. Alternatively the wall creeper is placed with the rather dumpy nuthatches and not in its own family. All are small with long, slender and down-curved bills, and all have long and strong toes with well-developed claws for climbing around on tree trunks and branches, or rock and cliff-faces in the case of the wall creeper. They are generally cryptically coloured except in flight, the most colourful being the wall creeper with much red, white and black clearly visible on the open wings.

Wall creepers are vagrants to North Africa, as is the Eurasian tree creeper, but the short-toed tree creeper has a resident population in the woodlands of far north-western Africa. Only the spotted creeper has a wide sub-Saharan distribution, also extending to India, but over its range population spread tends to be patchy. Woodland and forest constitute its primary habitat, where it forages for small invertebrate prey. In much the manner of many woodpeckers, this creeper starts its foraging low down in a tree and works its way, often in a spiral, to the top and then flies on to repeat the process in the next tree. Spotted creepers also resemble woodpeckers in the use of their tails as stabilising props on vertical trunks and branches while probing under bark, moss and lichens. The nest is placed on a branch: a small cup-shaped structure of plant material bound with spider web. ■ SEE PAGE 162.

Above and right: *Black cuckoo-shrikes, female (above) and male (right).* (Photos: Warwick Tarboton, ABPL)

Cuckoo-shrikes

FAMILY CAMPEPHAGIDAE

The family name, meaning "caterpillar-eaters", is more appropriate than the common name as these birds are not related to either the cuckoos or the shrikes. Of the 72 species of cuckoo-shrike and minivet in the world (occurring widely in Asia, Australia and Africa) only 11 reside in continental Africa.

Cuckoo-shrikes are small to medium-sized, have fairly stout bills that are slightly hooked at the tip, and nostrils that are completely or almost hidden by feathers and numerous short bristles. The legs are short but the feet are fairly strongly developed and the wings are quite long and pointed. Plumage is variable according to species but all share a common feature: the feathers on the lower back and rump are thick, with stiff and pointed shafts that are very loosely attached and partially erectile. It is surmised that this may function as a defence mechanism, because when threatened cuckoo-shrikes have been recorded to turn their back on the predator or other perceived threat. Most species are black, grey or white, or combinations of these, but a few are more brightly coloured, for example the blue cuckoo-shrike. In a few species the males and females look alike but in most they differ markedly in coloration and markings. In the genus *Campephaga* the males are glossy black, and some have patches of red or yellow on the shoulders. The oriole cuckoo-shrike takes its name from its similarities to the forest orioles both in coloration and size.

Caterpillars make up a very important component of the diet of cuckoo-shrikes but other insects and invertebrates are also eaten. The fairly widespread black cuckoo-shrike is known to include some fruits and seeds in its daily menu; some other species may do so too. They forage singly, in pairs or small parties, and the forest dwellers not infrequently join mixed-species flocks. Some species have very disjunct distributions but it is difficult to say whether this merely reflects a lack of records and the dense nature of their forest homes, or is a true indication of their range. The purple-throated cuckoo-shrike occurs in the tropics but in a few apparently discrete patches, in many cases linked by seemingly suitable habitat. A similar distribution pattern is apparent in the grey and the blue cuckoo-shrikes.

The breeding habits of several species is virtually unknown but in general the nest is a small, shallow cup made of plant material, commonly decorated with lichens and bound to a horizontal branch or tree fork with spider web. Between one and three eggs is the normal clutch size. When the chicks grow the nest appears ridiculously small. The nestlings of known species have a covering of snowy-white or greyish down, but before fledging they acquire plumage similar to that of the female. ■ SEE PAGE 162.

Drongos

FAMILY DICRURIDAE

Of the seven drongo species occurring in Africa, two are restricted to the Comoro Islands off Africa's Indian Ocean coastline. All Africa's drongos are entirely dressed in black and a sheen is usually distinguishable to their plumage. They are somewhat shrike-like in overall appearance and have stout, arched bills that are hooked at the tip, with short bristles around the nostrils. Most species have forked tails, with the obvious exception of the square-tailed drongo.

Drongos usually select prominent perches from which to launch their assaults on unwitting insects. Prey is frequently hawked in the air but ground pounces are common, with smaller prey taken back to the hunting seat. Larger prey such as locusts are dismembered with the bill before the edible parts are swallowed. We have watched fork-tails hawking insects disturbed by the hoofs of cattle and sheep, even on occasion using these beasts as mobile perches. Other drongo species may do this as well. They also frequent game herds, including those of elephant, buffalo and giraffe, and even ground hornbills. Fork-tailed drongos commonly hawk ahead of bush and grass fires, when several dozen may gather, and we have also seen them hunting at night in the light of powerful spotlights – not a bird to lose any opportunity! Small birds are also caught and eaten, and these may be carried to the perch in the feet.

Drongos are pugnacious and aggressive birds far beyond their size. It is said that certain tribes in the Congo call the velvet-mantled drongo the "angry leopard" because of their tenacity and "true grit". They will pursue and attack predators and raptors without hesitation, especially when breeding. These brave birds are also noisy, uttering a range of loud, harsh and grating calls when perched and on the wing.

The nests, thin-walled saucer-shaped or cup-shaped structures made of plant fibres and bound with spider web, are usually slung hammock-like in horizontal tree forks. Two or three eggs make up the normal drongo clutch, and the nest site is ferociously defended against threats or perceived threats, including humans. ■ SEE PAGE 162.

Left: *The fork-tailed drongo has an extensive sub-Saharan range.*

Above: *The square-tailed drongo.* (Photo: Richard du Toit)

Flycatchers

FAMILY MUSCICAPIDAE

Another large, mainly insectivorous group of birds are the flycatchers of the Old World. Africa is home to approximately 90 of these usually small, active and often attractive members of the avian horde. This family includes the beautiful silverbird, the monarchs, paradise flycatchers, wattle-eyes and the diverse batises.

In most species the bill is fairly broad and short, and somewhat flattened, and there are well-developed rictal bristles around the nostrils. These, as in the nightjars, increase the effectiveness of the "flytrap" mouth. The legs and feet are not particularly well developed, as flycatchers spend little time on the ground, most of their insect prey being plucked from the air or gleaned from trees and bushes. The wings are short and rounded in some species but longer and pointed in others. Tail length is also variable, from very short in some of the wattle-eyes to extremely long in male paradise flycatchers and monarchs. In the latter the tail may be double or more the length of head and body. We have seen these birds in many different locations but the tamest we have encountered are in the Dhofar of Oman, where a pair were using the backs and heads of camels as flycatching perches.

As a group the flycatchers have very variable plumage, with some being brightly attired, such as the silverbird, yellow-bellied wattle-eye and the monarchs, while many others are dressed in dull browns, greys and black and white. Although not always a hard and fast rule, in the dull or plain-coloured species sexes are similar but in the more brightly coloured and boldly marked forms sexes often differ. A few species have quite pleasant voices but many have harsh, monotonous or repetitive calls, no match for their thrush and warbler cousins.

The wattle-eyes, so called because of the fleshy, brightly coloured wattles above or around the eyes, represent a group known as the puff-backed flycatchers. When alarmed or displaying they raise and spread out their long rump feathers.

Most species are solitary foragers, although several do join mixed-species parties working through undergrowth or forest canopy. Some are sedentary, others local migrants and a few are long-distance travellers, but the majority are African endemics. One of the master migrants in this family is the spotted flycatcher, with large numbers leaving their Palaearctic breeding grounds in autumn. Birds ringed in Europe, as far north as Helsinki, have been recovered in South Africa where

Far left: *Female African paradise flycatcher bringing food to the young.* (Photo: John Carlyon)
Left: *Spotted flycatchers are non-breeding migrants to much of Africa, breeding only in the extreme northwest.* (Photo: John Carlyon)
Below right: *Chinspot batis.* (Photo: Richard du Toit)
Below left: *The black flycatcher is almost the size of the fork-tailed drongo, with which it frequently associates.* (Photo: John Carlyon)

this species occurs virtually throughout during the northern winter. Another long-distance specialist is the collared flycatcher, which "commutes" between Eurasia and Africa. A typical local migrant is the fairy flycatcher of southern Africa, which breeds in a narrow belt across South Africa, and then moves northwards into lower-lying areas. There are also those that have resident as well as migrant populations. A few montane-dwelling species undertake altitudinal movements in order to escape the seasons with inclement weather.

The majority of flycatchers build small, neat cup nests from plant material, with spider webs as binding, in a tree or bush. However, a few follow other architectural dictates. Livingstone's flycatcher constructs a ball of leaves bound together with spider webs, with a side entrance. A few nest in tree cavities but many African species' breeding habits and behaviour are poorly known. A small clutch of two or three eggs seems to be usual, but sometimes one or four may be laid, depending on the species. ■ SEE PAGE 165.

Far left: The African black-headed oriole feeds on insects, fruit and nectar.
Left: The dusky flycatcher is a bird of forest and dense woodland. (Photo: John Carlyon)

Orioles

FAMILY ORIOLIDAE

The name oriole derives from the Latin for golden or yellow, and referred originally to the widespread Eurasian bird now called the golden oriole – a fine language muddle, the Eurasian golden golden! The nine orioles occurring in Africa, and those in the rest of the Old World, are sometimes referred to as the true orioles, as opposed to the unrelated orioles of the New World. It is generally held that the true orioles are closely allied to the crow family, with which they share a number of anatomical features.

Orioles are the size of large starlings. Males have predominantly bright yellow plumage and in the majority of species a black head, but in two species just black eye-stripes. The name of the green-headed oriole is self-explanatory. The females of some species are similar to the males, but in others are duller and often with a more greenish hue. Bill colour is reddish to orange-brown, as are the eyes – altogether very attractive birds but unfortunately they are more often heard than seen. All compensate for their withdrawing nature by producing pleasant piping, bubbling calls that are one of the delights of the avian chorus.

These are birds of forest and woodland, as well as savanna with tree clumps. They have a strong undulating flight that brings to mind that of the woodpeckers. A few species are sedentary, others undertake local or intra-African seasonal movements, but the Eurasian golden oriole occurs on the continent only during the northern winter as a non-breeding migrant.

Orioles are largely solitary birds, or live in pairs, and do most of their foraging in trees. At least in a few species, individuals may join mixed-species foraging parties working through the dense undergrowth of forests. They are omnivorous, feeding on insects, including hairy caterpillars which are vigorously rubbed against branches, fruits, flowers and nectar.

Their nests are intricate cup-shaped structures constructed with plant material, including lichens, and hung in hammock fashion below a branch. The nest is usually located well away from the tree trunk. ■ SEE PAGE 162.

Ravens, crows and kin

FAMILY CORVIDAE

Ravens and crows appear frequently in literature and mythology, feared by some because of evil associations, beloved by others. Whether as companions of witches and wizards, birds of wisdom and knowledge, the friends and foes of man, members of the crow family feature more strongly in our psyche than any other passerines. They are familiar to us all: a pair of white-necked ravens circling their domain, a flock of fan-tailed ravens tumbling and weaving in updrafts against a cliff edge, a squadron of pied crows patrolling a road for the animal unfortunates that succumbed to our iron-clad steeds ...

The family includes the ravens and rooks, crows and jackdaws, choughs and jays, magpies and the piapiac. With just 17 species, of the world total of approximately 100, Africa is not well endowed with corvids but those that are present are mostly easily observed and some occur in substantial numbers. Only six species are African endemics, the others being shared with Eurasia. The Eurasian jay, magpie, Alpine chough, jackdaw, hooded crow and raven have but a toe-hold in the north of the African continent; brown-necked and fan-tailed ravens extend into the Horn and northern East Africa. House crows, introduced to East and South Africa, are perfect examples of the old adage, "Act in haste, repent at leisure." They have spread rapidly along coastlines where they concentrate around human settlements, and their breeding success is high. Efforts to eradicate them, for example on Unguja Island, Zanzibar, have been abject failures.

Members of this family are ardent scavengers, frequenting rubbish dumps, fishing camps, you name it – little goes unnoticed by these intelligent birds. With a few exceptions they are able to coexist with people and their ways. This coexistence may be peaceful, but at times there is conflict: the intelligent mammals pitting their skills against the birds with brains. This battle between man and bird has raged long and is still in progress, because despite the shooting, trapping and poisoning these birds continue to adapt to change; their ability to avoid succumbing to almost constant persecution is legendary.

This very successful passerine family, considered by some to be the most intelligent of all birds, has its roots in ancient avian evolutionary times. These passerines – we shall refrain from calling them songbirds as most have harsh cawing and croaking calls – are a rather unspecialised group of medium to large birds, including the largest of all perching birds. All have strong, well-developed bills that range from the slender, pointed, down-curved bill of the choughs to the massive one of the appropriately named thick-billed raven, an endemic of Ethiopia, Eritrea and Somaliland. All have well-developed and

Left: *The pied crow is widespread in sub-Saharan Africa, and is occasionally recorded in Libya and Algeria.* (Photo: John Carlyon)
Below: *The Alpine chough only occurs in the far north-west of the African continent.*

strong legs and feet, and on the ground they have a swaggering "dandy-like" walk. In crows, ravens, choughs and magpies the plumage is dominated by glossy blacks, white and on occasion grey; only the jays have colourful plumage. The bills and legs of most are dark to black, but those of choughs are red; the alpine chough sports red legs and a yellow bill.

Choughs are small, mountain or hill-country crows that are known for their impressive aerial acrobatics, including steep dives and somersaults executed with much wheeling.

The jays, of which only the Eurasian occurs marginally in Africa, are a large group of small crows that have a rather less omnivorous diet than their black cousins. Their diet includes a large percentage of vegetable matter and their distribution can be closely tied to oak trees bearing the acorns that form an important component of this. In general, however, corvids are true omnivores taking both animal and plant food; they are true survivors and opportunists. Some prey items can be quite large and seemingly unwieldy. We know of two "anvil-sites" used by white-necked ravens for smashing the carapaces of small tortoises; they are carried in the bill, or on occasion in the feet, to a height of 20 m and more, then dropped to smash on the rock below. This process is repeated until the raven can gain access to the flesh of the helpless reptile.

The magpie is also a jay. Its African distribution is limited to the north-west, and this bird is sometimes held to be a distinct species because of certain external features. Two of Africa's resident crows, the piapiac and Stresemann's bush crow, bear lit-tle resemblance to one's idea of a typical corvid. In West Africa the piapiac, with entirely black and brown plumage and a long graduated tail, is called the black magpie. This unusual crow extends in a broad belt from Mauritania across the Sahel to northern East Africa. Stresemann's bush crow is rather starling-like and has white to creamy-white and black plumage, with bluish bare skin around the eyes. It is restricted to bushed savanna in southern Ethiopia.

With the exception of the bush crow and the magpie, which construct large domed nests, the vast majority of corvids build bulky to fairly flimsy structures of twigs and other plant materials. Black and pied crows commonly build on telegraph poles and frequently incorporate wire, plastic and other human debris into their breeding platforms. Several other species are cliff-ledge nesters, but most construct their nurseries in trees. Some are social nesters, others nest in loose colonies, still others are solitary in their nesting habits. Most crows lay an average of four eggs, occasionally more depending on the species.

It is fitting to end off the description of the corvids as we began it: with a reference to literature:

... the female flies to the nest. The male follows — and there, and there only, does the sacred rite of mating take place. With the rook the nest is the marriage couch. Although the introductory actions may take place away from the nest, on it alone do the birds bring their courtship to its fulfilment. G.K. Yeates, *The Life of the Rook*

■ SEE PAGE 162.

Left: *Isolated populations of the chough occur in the high Atlas of Morocco and Algeria, and in the Ethiopian Highlands.*
Left middle: *In Africa the jackdaw is only found in Morocco and northern Algeria.*
Below: *Pied crows often gather in flocks.*

Above: *The house crow occurs in scattered populations along the Red Sea coast and on the Indian Ocean coastal plain. It has become a serious pest in some areas, such as Unguja (Zanzibar).*
Left: *The white-necked raven has a heavy and well-developed bill.* (Photo: John Carlyon)

Shrikes and tchagras

FAMILIES LANIIDAE, MALACONOTIDAE AND PRIONOPIDAE

The shrikes, helmet-shrikes, bush shrikes, tchagras and their kin are very well represented in Africa, with a total of 79 species in three families (some say only one) and 12 genera. The tropics of Africa are the centre of shrike development and abundance; it is deemed likely that this family developed here and spread from this continent into Eurasia. These are the "mini-raptors" of the passerine world, pugnacious, dashing and bold. Some are tagged with the label of butcher-bird or jacky-hangman, a reference to the way in which they impale their prey in larder-fashion on barbed wire or thorns.

These small to medium-sized passerines are armed with stout, toothed and sharply hooked bills. Although equipped with strong and well-developed legs and feet, these are not used to kill prey as in most of the true raptors. In most species the wings are fairly short and rounded, and the tail is usually moderately long. Some shrikes sit in prominent positions watching for prey, others skulk among the undergrowth when foraging. The aggressive hunting shrikes, such as the great grey and common fiscal, are disliked by many gardeners and bird-lovers because of their forays against such prey as small birds, reptiles and the like. Few may feel for that grasshopper or beetle impaled on the thorn but let it be a tiny white-eye or baby chameleon and the hatred rises. The strategy of the perch-hunters is to watch, wait and then pounce. Larger prey, once killed or subdued, may be carried in the feet to the dismembering site but as mentioned above the claws, or "talons", play no part in killing the unfortunate victim. Many shrikes are surplus-killers, that is they kill more than for their immediate

needs, and several carcasses may be impaled for later consumption. However, often the shrike does not return to his larder and the carcasses dry out. Having examined many such larders we cannot help but wonder at the strength of these birds when one sees hard-carapaced beetles impaled. A few of the bush shrikes include, on occasion, some wild fruits and berries in their diet.

Plumage coloration in many species is dominated by black, white, greys and browns, and in the bush shrikes greens and yellows. A few of the bush shrikes and the gonolek have their underparts entirely red, a few merely have red throat-patches. Most have harsh, grating calls but the characteristic calls of the bush shrikes are not unpleasant to the human ear. Several species, such as the fiscal shrike, are good mimics of the calls of other birds but what purpose this serves is not known.

The helmet-shrikes are boldly marked and have distinctive erectile crests of differing length; probably the most impressive is the all-black, yellow-crested helmet-shrike. Each species has a distinctive wattle around the eyes. They differ from other shrikes in being more social and less aggressive, and more foraging in their insect-hunting behaviour, even during the breeding season. At least in several if not all species, helpers assist the parent birds in nest building and feeding the young. Shrike nests vary but are always open cup- or bowl-shaped structures made of plant material; some are relatively neat, a few rather untidy. Most shrikes have short to medium-length tails but in the long-tailed fiscal and especially the magpie shrike the tail is long. ■ SEE PAGE 165 & 166.

Above left: *Common fiscals are aggressive hunters and kill birds up to the size of palm doves, but this is unusual.* (Photo: John Carlyon)
Above: *The barbary shrike from west Africa.*
Left: *Slate-coloured boubou, a bush shrike.* (Photo: Duncan Butchart)
Far left: *The magpie, or long-tailed, shrike is not easily mistaken for any other species.*

Starlings

FAMILY STURNIDAE

One of the world's best known starlings is the speckled, short-tailed common starling, misguidedly introduced into many countries beyond its European range. It is an aggressive and problematic settler that comes into conflict with humans at almost every turn. The common mynah is another unfortunate transplant, an aggressive invader that has settled well into a number of coastal and inland towns in eastern and southern Africa. Creating urban mess and rural competition while displacing indigenous birds, these noisy species have blighted people's attitudes to the amazingly diverse starling family. It is unfortunate that starlings gain a bad reputation because of these two man-aided avian invaders.

The starling family is diverse, with some 50 species in Africa, including a number of the real gems of the African avian world. The aptly named superb starling, golden-breasted starling, plum-coloured starling and the numerous species of glossy starling are among the most colourful and striking the continent has to offer. Almost half of the world's starling species occur in Africa, most as endemic residents. The origins of starlings are poorly understood, but some believe that they stem from primitive thrush-like stock. Their centres of greatest development and diversity lie in Africa and Asia.

Most starling species are noisy and active birds of moderate size with fairly long, usually straight, bills. Their legs and feet are well developed and strong, a sign that they spend much of their time on the ground foraging. They are walkers and runners, seldom hopping as a means of locomotion, and in the main they are strong fliers. A few species have long tails, such as the bristle-crowned, white-crowned, golden-breasted and ashy starlings, but the majority have tails of moderate length. In most species the male and female look alike, with eye-catching plumage. The shimmering blues, greens and purples of the aptly named glossy starlings are made even more dramatic by their habit of moving in fairly tight-knit flocks, sunlight enhancing this avian colour-burst. The "red-winged" starlings also have a sheen to their dark plumage but in their case the extensive redness of the wings in flight is the impressive feature. A rather drab and nondescript bird, for much of the year, is the widespread wattled starling. During the breeding season the male develops black wattles that contrast strongly with bare yellow skin on the head.

Most species are solitary nesters but a few are colonial, one of the most dramatic being the wattled starling. These birds build untidy balls of sticks, with the nests often touching each other, in trees or bushes, and some colonies may include hundreds of breeding pairs. Then there are the natural tree-hole nesters such as the glossy starlings, the pied starling excavates burrows in earth banks, the red-wings nest mainly in rock crevices and nests of the common starling are often built under the eaves of buildings or in other man-made structures. In most species, at least those that have been studied, the bond between male and female is strong, and both share duties at the nest. Clutch size is variable in a range of two to nine eggs but between three and five is more usual.

Overall the starlings are omnivores, but some species take more animal food, others favour berries, fruits and seeds; several take nectar from aloes and other flowers. Golden-breasted starlings have a taste for termites, as do many other species, and these are often dug out with side-flicks of the bill. Many starlings do much of their foraging on the ground but readily take to the trees and bushes to pick berries. Starlings are opportunists when it comes to foraging, grasping any opportunity for a meal. During a severe drought in Namibia aardvark, normally nocturnal mammals, were forced to excavate for termites throughout the day. As we watched and photographed the aardvark, each (we encountered three) was accompanied by noisy flocks of glossy starlings, the birds dashing in to snatch the insects as they were exposed. ■ SEE PAGE 166.

Heuglin's robin-chat at the nest. (Photo: John Carlyon)

Thrushes, robins and chats

FAMILY TURDIDAE

This large family is very well represented in Africa, with about 136 species. Some authorities recognise only 125 species but even this number represents more than a third of the world's tally. In all there are 26 genera, ranging from a single species to the 20 members of the *Oenanthe* (wheatears). Its ranks include many of the finest songsters in the avian world. They bring to mind the Gaelic proverb: *Deserted indeed is that country where no voice of bird is heard.* And how many poems have been written about the nightingale, its song having inspired great and not so great writers to put quill or pencil to paper? The words of Izaak Walton, in his masterpiece *The Compleat Angler*, sum up the beauty of the nightingale's song:

Lord what music hast thou provided for the Saints in heaven, when thou affordest bad men such music on Earth!

Any who have listened to the melody of song thrush, the wake-up call of Heuglin's robin-chat or the rather sad and haunting song of the European robin cannot help but pause and absorb. Several species are excellent mimics, including the talented red-capped robin-chat. We have heard them uttering near-correct calls of the African fish eagle, just faltering on the last note or two; we once set about hunting down a cat in a nature reserve only to return red-faced and admit we had been tracking none other than this robin-chat!

Included in this family are the thrushes, robins and robin-chats, wheatears, chats, akalats and alethes. They range in size from small to middling. Most species come in shades of brown, grey or rufous; some have distinctive breast speckling, others distinctive red, orange or yellow underparts. Many have black and white head markings, a few have colourful throat patches.

Most species are terrestrial in their foraging, some arboreal, and the different species have succeeded in occupying virtually every African habitat. Many are forest and woodland dwellers, others occupy the grasslands, deserts and mountains, and several have settled in easily with people. In our garden we have a pair of Karoo robins, rather dull-looking birds, but this is more than made up for by their character. Like the European robin to the north, these denizens of the Karoo rush in between one's feet and spade, snatching up tasty insect morsels.

It is believed that the thrush family is closely allied to the Old World warblers and flycatchers, although this is in some dispute. The *Turdus* genus contains the largest members, such as the well-known olive thrush, Eurasian blackbird and groundscraper thrush. Distinctive and more easily seen, because of their more exposed habitat, are the rock thrushes of the genus *Monticola*, of which eight species occur in Africa. They are medium-sized thrushes and most have greater or lesser amounts of blue in the plumage of the males. The exceptions to a life in forest and woodland are most clearly the wheatears and many of the chats, most of which occupy open or lightly scrubbed terrain. Many of these birds have an almost exaggerated upright stance, frequently standing on some prominent feature such as boulder or termite mound. The wheatears, which by the way are not named for any association with ears of wheat, but for the fruity old Anglo-Saxon for the rather inelegant "white-arse", are nearly all dressed in combinations of black and white and fawns. In fact in several species – confusingly different subspecies or races have different combinations of black and white and could be mistaken for different species.

As in all large avian families there are always those species that have very restricted and localised ranges, as well as those that have extensive distributions. Many are African residents, a few are intra-African migrants, and those such as the bluethroat, pied wheatear, fieldfare and song thrush enter the continent as non-breeding migrants. A few species breed in North Africa at the edge of their Palaearctic range, including that brilliant songster, the nightingale, also the black redstart, redstart, northern wheatear and rufous-tailed rock thrush. Species such as Swynnerton's robin have strangely disjunct ranges with four apparently separate populations from southern into East Africa, each located in highland forest areas. In fact many species of the thrush family, particularly in the tropics, have a disjointed known range, a patch here and a patch

there. Another species is the east coast akalat occurring in three tiny populations on and close to the eastern seaboard. Then there is the Gabela akalat which is apparently restricted to an area of less than 1 000 km² in west-central Angola, and the Usambara akalat which is only known from the western Usambara range in north-eastern Tanzania. Needless to say, with many species we have little understanding of their requirements, biology or anything else.

The thrush family have members that are almost pure insectivores, but many are omnivores which eat both animal or plant food, mainly berries, fruits and seeds. A few are more specialised, such as the red-tailed ant thrush, moving ahead of columns of the dreaded army, or soldier, ants plucking up small arthropods fleeing the ravenous horde. Some, including the boulder chat, will take small vertebrates such as lizards, and the collared palm thrush is known to prey on frogs.

Most species construct a typical "bird's nest": a simple cup with a varying amount of plant material, often in a tree fork, but in many chats and wheatears on the ground. Several species, in particular the three palm thrushes, use mud as the principal construction medium. A few species build their nests either on the ground or above the ground, this no doubt being largely dependent on suitable sites. The ant-eater chats excavate short burrows in suitable substrate, often inside the walls of such large burrows as those made by aardvark and porcupine. In most species clutches are fairly small, from one to five eggs. ■ SEE PAGE 163.

Far left: *Capped wheatears are one of Africa's most widespread.* (Photo: John Carlyon)
Left: *Arnot's chat is a species of well-developed woodland.* (Photo: John Carlyon)
Below right: *The buff-streaked chat is a southern African endemic.* (Photo: John Carlyon)
Below: *Sooty chat, Uganda.* (Photo: Duncan Butchart)
Below far left: *Female Cape rock thrushes are not as brightly coloured as the males.* (Photo: John Carlyon)

Tits, titmice and penduline tits

FAMILIES PARIDAE AND REMICIDAE

In Africa and Eurasia we call these birds tits but our American cousins favour titmice and chickadees, seemingly a hangover from past centuries. The vernacular names derive from Anglo-Saxon, "tit" meaning a small or tiny object, and the "mouse" part apparently being a corruption of "mase", a name applied to several small birds.

On the African continent 21 species of tit occur (family Paridae); most are endemic but a few found only in North Africa are at the edge of their Palaearctic range here. The penduline tits (family Remicidae) are sometimes included with the Paridae. The penduline tits take their name from the nature of their nests. They are smaller even than the tits, or titmice. Generally dull in colour, they usually lack the distinctive head markings of their larger cousins.

In Africa the tits have never become as tolerant of humans as those occurring elsewhere, particularly the blue tit, which people welcome to their gardens and bird-tables with bags of peanuts and the like.

Tits are all tiny birds with soft, thick plumage. Most are mainly black, white or brown, and many have distinctive markings on the head. Some, such as blue and great tits, have the underparts yellow, the former with powder-blue on the wings, tail and top of the head. Their bills are short and stout, and the legs and feet are quite small but strong — important as tits spend much time clambering around in trees and bushes, frequently hanging upside down when feeding. Although most species have short tails, the long-tailed tit, a vagrant to North Africa, is an exception as its name makes clear.

Given their small size it is hard to credit that tits are believed to be closely related to the much larger crows, but in part this is ascribed to their intelligence and learning skills. Tits forage in pairs but commonly form small parties, not infrequently joining mixed-species flocks as they move restlessly through the trees and bush thickets. Insects and other small invertebrates such as spiders make up the bulk of their food but certainly in some of the northern hemisphere species small fruits and seeds are also taken. At least in some species, possibly all, hard-bodied insects and seeds are held in the foot and tapped with the bill, almost parrot-fashion.

Most nest in natural tree-holes, but they also nest in rock crevices and in holes in rock walls, and even hollow gate- and fence-posts. The penduline tits, of which eight species occur in Africa, include some of the avian world's finest nest-builders. Most construct hanging, bag-like nests of felted fluffy plant and animal fibres, completely closed with a short entrance tube that is pulled shut by the adults. Below the entrance tube there is a small hollow or ledge that acts both as a perch and a false entrance. ■ SEE PAGE 162.

The blue tit has a wide European distribution but in Africa is found only along the north-west African Mediterranean seaboard, with an apparently isolated population in north-eastern Libya.

Waxwings

FAMILY BOMBYCILLIDAE

The Bohemian waxwing occurs only as a very rare vagrant in North Africa. The grey hypocolius has been recorded only as a rare vagrant to Ethiopia in 1850, and several times near the Sudanese-Egyptian border. The fact that it occurs as a regular migrant to eastern Arabia indicates that it may be expected on the continent from time to time. ■ SEE PAGE 163.

Top left: The Abyssinian white-eye occurs extensively in East Africa, penetrating Arabia as far as the Dhofar region of Oman.
Above: The taxonomy of the white-eyes is complex; here a Cape white eye.
Left: The nest of the Cape penduline tit is constructed of tightly felted woolly animal and plant fibres.
Far left: Penduline tit building its nest.
(Photo: Clem Haagner ABPL)

White-eyes and speirops

FAMILY ZOSTEROPIDAE

Thirteen species of white-eyes and speirops occur in Africa, of which five are restricted to islands within the Afrotropics. White-eyes range eastwards through Asia and have successfully occupied many oceanic islands. Several species have a host of races, many of which cannot be separated in the field. All are tiny at considerably less than 20 g. Plumage is yellowish-green, and a ring of white encircles the eyes. Some species show varying amounts of grey, brown and white, usually on the belly or chest. They have fairly short, slender and pointed bills and the tip of the tongue is equipped with a brush-like structure.

When they are not breeding white-eyes are gregarious, foraging in trees and bushes, flitting back and forth and seemingly constantly on the move. Usually found in forest and different woodland associations, they commonly occur in suburbia and take up residence in terrestrial "islands" surrounded by unsuitable habitat. They are present in many small villages where there are trees and well-developed gardens.

These tiny birds feed on a mix of nectar, fruit, buds and small insects, and commonly forage in the company of other small birds. When nesting each pair establishes its territory, and a tiny open cup is constructed from plant and animal fibres and bound with spider webs. The nest is normally suspended from a horizontal fork in a tree or bush. Two to four eggs are laid and the young are fed by both parents. ■ SEE PAGE 167.

12

Specialised feeders

The two known African oxpeckers, a family unique to the continent, have evolved in such a way that they are entirely reliant on large herbivores to provide them with their "daily bread", in the form of ticks, insects and the wound tissue of their hosts. The benefits are not just one-sided, as the birds serve as "early-warning devices" for their hosts, signalling the approach of predators and other potential threats. The other specialist feeders are the nectar-eaters: the host of sunbird species and the two sugarbirds. Taxonomic issues aside, all to a greater or lesser extent feed on flower nectar, with varying quantities of small insects and spiders being eaten. As nectar feeders many are important pollinators of a wide range of flowering plants. The sunbirds include some of Africa's most brilliantly plumaged birds, the continent's "living jewels". They are all tiny (most weigh less than 15 g), with long, slender and often down-curved bills.

Oxpeckers

FAMILY BUPHAGIDAE

Oxpeckers are often considered to be highly specialised starlings, but because of their unique life style others prefer to place them in their own family. There are two species, the red-billed and yellow-billed oxpeckers. A possible third species has been reported but is as yet undescribed. This small oxpecker has been sighted attending buffalo in the Tai Forest of south-western Ivory Coast in West Africa. Sadly, it is deemed highly likely that as its principal host, the buffalo, is close to extinction here, this undescribed bird could well follow it over the brink without ever becoming scientifically known.

The two known species have very similar plumage but differ in the colour around the eyes, and obviously their bills. Much of an oxpecker's life is spent on large game animals and cattle, feeding on ticks, insects and wound tissue. The two species differ in their method of removing ticks: the yellow-bill plucks these parasites off, whereas the red-bill uses a sideways scissor-action. They are equipped with sharply curved and pointed claws to cling to their host, and the tail is stiff and pointed for use as a prop.

They normally feed in small groups, moving from host to host, but circumstances sometimes bring larger numbers together in temporary feeding aggregations. For example, where hippopotamuses are abundant such as along the Luangwa River, several small flocks attend these amphibious giants. When they are in the water the birds patrol their heads, probing the ears, nostrils and around the eyes; if the mammal submerges they move on to the next head. Apart from benefiting their hosts as parasite removers, oxpeckers are also very alert and warn the mammals of approaching danger. In areas with commercial cattle ranching and therefore dipping of the livestock in poison, oxpeckers have largely disappeared. ■ SEE PAGE 166.

Left: *A male orange-breasted sunbird.* (Photo: Lanz Von Horsten, ABPL)
Above: *The collared sunbird is one of the most widely distributed African sunbirds.* (Photos: John Carlyon)

BIRDS USING MAMMALS AS FEEDING STATIONS

In Africa more than 80 bird species, belonging to 32 different families, have been recorded as using mammals as part of their feeding strategies. In the case of the two oxpeckers, the yellow-billed (*Buphagus africanus*) and the red-billed (*Buphagus erythrorhynchus*), the birds are entirely reliant on mammals for their primary source of food. Bird and mammal feeding associations can be broadly divided into two categories: those where the bird and mammal eat the same general food, and those where the bird uses the mammal to help it obtain food that only the bird requires.

The classic case in the latter category is that of cattle egrets following cattle, as well as game species, gleaning insects and on occasion small vertebrates disturbed by the animals' hoofs. In fact the vast majority of bird and mammal feeding associations involve the birds as predators of mobile prey. It is a common sight to see such species as wattled, pied and glossy starlings perched on the backs of sheep and cattle, where they may on occasion take ectoparasites but more commonly rely on this mobile perch to disturb insects, which they then hop off to catch. We have watched red-winged starlings (*Onychognathus morio*) gleaning ticks from around the ears of klipspringer, and in south-western Arabia Tristram's grackle (*Onychognathus tristramii*) commonly associates with camels, both removing ticks and catching disturbed insects.

One of the most exciting feeding associations we have observed was on the northern fringe of the Jozani Forest on Unguja, Zanzibar. A small troop of Zanzibar red colobus monkeys (*Procolobus kirkii*) were feeding some 2 m above the ground in a bush that was heavily in berry, with an Ader's duiker (*Cephalophus adersi*) and a flock of crested guineafowl (*Guttera pucherani*) feeding on the fruits falling to the ground. Although many bird-mammal feeding associations are common and habitually repeated, some are rare and seldom observed. During a period of severe drought in northern Namibia we were undertaking survey work on a game farm. Here, over a period of several days, we observed aardvark (*Orycteropus afer*) excavating for their termite prey during the daylight hours. Each aardvark we encountered was accompanied by small flocks of common glossy starlings (*Lamprotornis nitens*) which rushed in to snatch termites whenever their mammalian companion dug into a termite colony. A male stonechat (*Saxicola torquata*) in our home village uses the backs of penned pigs to catch flies to great effect!

Far left: *Yellow-billed oxpecker on square-lipped rhinoceros.*
Middle: *Gurney's sugarbird occupies areas with aloes, proteas and other suitable nectar-bearing plants.* (Photo: John Carlyon)
Left: *The Cape sugarbird, confined to the fynbos of the South African coastal mountains, is easy to identify with its long, showy tail.* (Photo: Nigel Dennis, ABPL)

Sugarbirds

FAMILY NECTARINIIDAE

The two sugarbirds, like many others, are the subject of taxonomic wrangling: do they deserve their own family ranking (family Promeropidae), should they be placed with the sunbirds, or are they more closely allied to the thrushes?

Their overall dull plumage is brightened by yellow vent feathers and the particularly long tail of the males (although not as long in Gurney's sugarbird). They also have long, sharp-pointed, down-curved bills used for extracting nectar from protea flowers, which are essential to their survival. In the process they inadvertently transfer pollen from flower head to flower head. Protea flowers are teeming with a whole host of small insects and these are also eaten by the sugarbirds. Larger insects, such as those caught by hawking, are taken to a perch where they are beaten in much the same way as the bee-eaters "soften up" their prey; in fact at one stage they were thought to be closely related to that group of birds. ■ SEE PAGE 166.

Sunbirds

FAMILY NECTARINIIDAE

Sunbirds probably form part of the same family (the "nectar-feeders") as sugarbirds. There are 83 or so of these "living jewels" in Africa, and on this continent they reach their greatest diversity, particularly in the tropics. They extend eastwards to Southeast Asia and occur on a few islands in the Pacific Ocean, with just one species in north-eastern Australia. They are primarily sedentary but some localised nomadic movements may take place between feeding grounds. Although several may gather at suitable flowering plants, such as stands of aloe, sunbirds are primarily solitary or in pairs.

Most sunbirds are brilliantly coloured, as some of their names attest: malachite, orange-breasted, scarlet-chested, splendid, superb, regal, amethyst and scarlet-tufted. Several are rather dull little chaps, and females are clad in drab greys, browns and olive, but all have the characteristic thin, pointed and down-curved bill. The bill varies in length from species to species, an indication of their feeding habits. When feeding they may sip nectar directly from the flower, but when they cannot gain access in this way because the flowers are too large or have long tubular corollas, they pierce the petals near the base. Sunbirds have tongues that have evolved to aid their nectar feeding: the outer two thirds form a double tube which opens into a groove or trough to carry the nectar into the oesophagus. Most sunbirds perch when feeding but they also hover next to the flower. Some species, such as the scarlet-chested sunbird, hover more frequently than others. Sunbirds are also avid eaters of small insects and other invertebrates such as spiders. A few species may occasionally supplement their diet with small berries and seeds.

Most of these tiny birds (less than 15 g), have short to moderately long tails, but the males of a few species have greatly lengthened central tail feathers. These include the malachite, the two pygmy sunbirds, the purple-breasted and the golden-winged sunbirds. The males of some species have impressive courtship displays. Shall we ever forget watching a male scarlet-tufted malachite sunbird with wings spread and fluttering, and scarlet tufts open, calling and working his way up and down a giant lobelia flower head in the Aberdares, toiling away at attracting a mate. None of the African sunbirds are good musicians, except to their mates of course.

An outstanding feature of these birds is their nest-building abilities. The nests are hanging, closed purses with a side entrance near the top, constructed from plant fibres, including lichens, and usually bound together with spider webs. Although some nests are tidy and neat structures, in general they are rather untidy efforts. Some species construct an overhang, or porch, over the entrance to the nest. Most sunbirds have clutches of only two or three eggs. ■ SEE PAGE 166.

Above left: *Nest of a collared sunbird in a deep coral rag cave on Unguja.*
Above right: *Orange-breasted sunbirds are only found in the fynbos of southernmost South Africa.*
Left: *Sunbirds, here a male lesser double-collared, have long slender bills and tongues to gain access to nectar deep in even the largest protea flowers.*

13

Seed-eaters

Most birds covered in this chapter eat mainly seeds, but they often supplement this with plant buds, blossoms, fruits and even insects. Some are fine songsters, including one of the commonest birds around our Karoo home, the white-throated canary. In this throng we find some of the most accomplished of nest builders, the weavers. The sociable weavers of the south-west construct the largest nests, cooperating to build mighty roofed villages in trees or even on telephone poles. Although many seed-eaters are solitary nesters, some prefer a colonial way of life. This latter is taken to its extreme in the high-density cluster-nesting of the red-billed quelea, that scourge of grain farmers. The waxbills, mannikins, twinspots and fire-finches are all tiny but often brightly or distinctively marked and coloured. It is the sad lot of many seed-eaters to end up in cages well away from their homeland, in some cases because of their musicality, in others because of their plumage. Some suffer an even worse fate: death by fire or poison! Because of their chosen diet of seeds some species, in particular the quelea, come into direct conflict with the interests of man. For this temerity they must die.

In parts of South Africa certain weavers and bishops form into large flocks that descend on the grainlands, drawing the ire and revenge of farmers.

Canaries and buntings

FAMILY FRINGILLIDAE

These birds have the typical short, stout and conical bill evolved for seed-eating. Their wing shape is varied, from rounded to pointed; tails are usually of moderate length and in some species they are shallowly forked. Plumage coloration varies among the many species, although a great number of species are predominantly shades of brown, grey and yellow, and many are heavily streaked on the back or breast. Many buntings have the head heavily streaked in black and white.

Left: *Village weaver trying to attract females with fluttering wing display.*
Above: *Red bishop male in breeding finery.*

The 85 species in Africa include many endemics. Several are widespread, others are very localised, and a number are at the edge of their Palaearctic range in North Africa. Within this latter group falls the chaffinch, but on the Canary Islands off Africa's north-west coast we find the endemic blue chaffinch, where it is found mainly in montane pine forests, its only home. The canary for whom these islands are named, that much modified cagebird and brilliant songster, occurs naturally on several of the eastern Atlantic islands. Also of marginal occurrence in northernmost Africa are such seed-eaters as the serin, linnet, siskin, goldfinch, greenfinch, hawfinch, crimson-winged finch, crossbill and corn bunting, many of which breed in this part of the continent. A few others are rare vagrants: they inadvertently make African landfall after severe northern winters, or are just odd individuals blown off course.

Several members of this family are among the finest of avian songsters, in particular the canaries. Depending on the species, most are gregarious outside the breeding season, usually forming small flocks and not infrequently joining mixed-species flocks. We seem to keep coming back to the view through my office window but now in the winter months it is not unusual to see Cape, white-throated and black-headed canaries feeding in loose association – there are always pleasant distractions to draw me away from the keyboard ...

Members of this family occupy virtually every available habitat, from forest to near-desert, woodland to mountains; a very successful group indeed. Most are primarily seed-eaters but several take other plant parts as well, including fruits, seeds and nectar. Others are more specialised, such as the goldfinch which feeds on the seed-heads of thistles, or the crossbill which specialises in removing seeds from pine cones. Although several species are largely arboreal foragers, many glean their food from the ground.

All build cup-shaped nests, and depending on the species they are located in trees or bushes, or in a few cases on the ground. Two to five eggs make up the usual clutch. In many species the nests are cunningly hidden and are often difficult to find. ■ SEE PAGE 168.

Above: *Brimstone, or bully, canaries feeding nestlings.* (Photo: John Carlyon)
Right: *Yellow-fronted, or yellow-eyed, canaries have a very wide range in sub-Saharan Africa.* (Photo: John Carlyon)
Far right: *The Cape bunting often frequents human settlements and can become very trusting.* (Photo: John Carlyon)

Sparrows, weavers, bishops, widows and queleas

FAMILIES PLOCEIDAE & PASSERIDAE

This large group of birds counts some 135 species on the African continent, falling into 21 genera, the most numerous being the weavers of the genus *Ploceus*. These birds constitute some of the most familiar to humans; not always viewed, however, in a friendly manner. Here we find that cosmopolitan urbanite the house sparrow, the dreaded grain-consuming quelea hordes and some of Africa's finest architects, the appropriately named weavers. There are also those few that are barely known, some from but a handful of museum specimens; still others are so rare or live in terrain so difficult of access that they are seldom sighted, never mind studied.

The members of this family, or families if one prefers, share a number of common features. All are small to medium in size, have stout, conical bills adapted for seed-eating and are flocking birds and colonial nesters. There are of course a few exceptions: some forest inhabitants are more solitary. Plumage coloration is highly variable, although many of the weavers are predominantly yellow, and in breeding attire the long-tailed male widows are largely black, usually with a splash of colour on the shoulder, wing or head. The bishops likewise have bright red or yellow plumage set against black. But once the breeding season is over the males of most species moult and take on the drab browns that the females carry throughout the year.

Although seeds form the mainstay of their diet, most include some insects and fruits on their menu. A few, especially some of the weavers, also take nectar whenever it is available.

The weavers are far better represented in Africa than in any other part of the world and it is generally believed that they originated in Africa. These birds can be broadly divided into three groups: the buffalo-weavers, sparrow-weavers and typical weavers. The buffalo-weavers are among the largest in the group and they construct bulky, untidy stick nests that may have two or more inner chambers, each occupied by a pair of birds. Several of these structures may be located in a single tree. Another truly colonial nester, and one of the greatest architects and builders among the feather-folk, has to be the sociable weaver of the dry west of southern Africa. Usually constructed in large trees but also on telegraph poles, these nests reach massive proportions, and their weight can break even heavy tree limbs. Larger nests may be as much as 4 m deep and 7 m across, and construction is mainly with dry grass stalks, roofed over with twigs and sticks. The nest chambers, from five to 50 per nest, are entered from below via a tunnel that may be as much as 25 cm long. It is not only the builders who breed or shelter in their nests but pygmy falcons, rosy-faced lovebirds and red-headed finches make extensive use of them.

Far left: *The true weavers produce some of the most complex nests in the avian world.*
Left: *The male Cape sparrow is much more boldly marked than the female.*
Below: *Female Cape sparrows, or mossies, bathing.*

Many sparrows build round, untidy nests that usually have a side entrance, but a few nest in tree-holes or rock crevices.

It is without doubt the so-called true weavers that construct the most intricate nests. Most are noisy colonial nesters, constantly chattering with much toing and froing. The nests are generally constructed from grass blades or shreds torn from broader leaves, which are carefully woven and attached to a branch, twig or reeds. In many species it is possible to identify the bird by looking at the nest. Most nests are flask- or retort-shaped, with a round nest chamber, and may or may not have an entrance tunnel. Species such as the forest weaver weave particularly long entrance tunnels. In many colonial-nesting species males may construct several nests in a season, attracting females by hanging below the nest, calling and variously fluttering the wings, rocking or swinging from side to side.

Bishops also construct woven nests slung between reeds but the male of the red bishop has evolved a spectacular courtship display. In front of his harem of two or three females he undertakes a slow bouncing flight, clapping his wings at brief intervals and puffing out his brilliant red and black plumage. He also sits puffed up on a reed stem, calling and looking twice his normal size. Many male weavers are polygamous, but without doubt the most prolific of all breeders is the red-billed quelea, the bird that lives in huge flocks. Breeding colonies can cover several hectares and a single tree may house hundreds of nests. Efforts to fire-bomb and poison these colonies, because of their depredations on cereal crops, have little overall impact on quelea numbers. Large non-breeding flocks of red bishops and certain weavers also draw the ire of farmers when hundreds descend on commercial crops. ■ SEE PAGE 167.

Below: *Grosbeak weavers construct very neat domed nests.*
Bottom: *Part of a large village weaver colony.*

Top: *Vieillot's black weaver, Uganda.* (Photo: Duncan Butchart)
Middle: *Large flocks with thousands of red-billed queleas turn and wheel as one.*
Above: *Red-billed queleas in non-breeding plumage. This tiny bird sometimes gathers in flocks hundreds of thousands strong.*

Far left top: *African, or blue-billed, firefinch female. This tiny finch occurs widely in sub-Saharan Africa.* (Photo: John Carlyon)
Top right: *Green-winged pytilia.*
Above: *The chaffinch occurs naturally as a resident and non-breeding migrant only in northernmost Africa. A small population near Cape Town stems from introduced birds.* (Photo: Penny Meakin)
Left: *The blue-breasted cordonbleu is fairly common but subject to local movement.* (Photo: John Carlyon)

Waxbills, mannikins, twinspots and firefinches

FAMILY ESTRILDIDAE

There are 81 species of these small birds in Africa, the vast majority being continental endemics. Many are brightly coloured and boldly patterned, with short, conical bills and short, rounded wings.

Out of the breeding season nearly all species are gregarious but they are solitary nesters, or in a few species nests may be located in loose groupings. Nests are round or oval balls of dry grass, with a side entrance tube, and depending on the species may be in a tree, bush, tree-hole or on the ground, in the latter case often well camouflaged. All lay white eggs, and the clutch size is usually in the range of two to seven. On hatching the chicks are characteristically covered in down. There are a few species, including the orange-breasted waxbill, that do not build their own nests, or do so only on occasion; instead they take over the nests of weavers, bishops or prinias.

Many of these attractive birds are caught for the bird-fanciers market, in fact many tens of thousands are caught and shipped from Africa each year. One of our favourites is the African quail-finch, a tiny seed-eater that is entirely terrestrial and in many ways is similar to the small gamebird from which it takes its name. It is not unusual in this family for mixed flocks of two or more species to forage together.

Although a few species are associated with forests and dense woodland, many occupy different types of savanna and marshland, and a few extend into dryland country. The vast majority do much of their foraging on the ground where they harvest the seeds, mostly of grasses, that are their staple. A few, such as the blue-billed firefinch, also snack on insects but more as a supplement than as a replacement for seeds. Many members of this family feed their chicks on insects. ■ SEE PAGE 167.

Suggested reading

A vast body of information on African birds can be found in the scientific literature but it is beyond the scope of this book to list such sources here. There are also many field guides and books dealing with the avifauna of different subregions, including a number that are long out of press. The following is just a short collection of fairly general works.

The birds of Africa. 5 volumes; various editors and authors. 1982-97. Academic Press, London.

Collar, N.J. and Stuart, S.N. 1985. *Threatened birds of Africa and related islands*. ICBP/IUCN, Cambridge; 3rd edition.

Cramp, S. *et al*. 1977-94. *Birds of the Western Palaearctic*. Oxford University Press, Oxford.

Dowsett, R.J. *et al*. 1993. *Checklist of birds of the Afrotropical and Malagasy regions*. Tauraco Press, Liége, Belgium.

Gosler, A. 1991. *The Hamlyn photographic guide to birds of the world*. Hamlyn, London.

Hall, B.P. and Moreau, R.E. 1970. *An atlas of speciation in African passerine birds*. Trustees of the British Museum (Natural History), London.

Hancock, J. and Kushlan, J. 1984. *The herons handbook*. Croom Helm, London.

Harrison, J.A., Allan, D.G., Underhill, L.G., Herremans, M., Tree, A.J., Parker, V. and Brown, C.J. (eds) 1997. *The atlas of southern African birds*. 2 volumes. BirdLife South Africa, Johannesburg.

Harrison, P. 1983. *Seabirds: an identification guide*. Croom Helm, London.

Heinzel, H. *et al*. 1995. *Birds of Britain and Europe with North Africa and the Middle East*. Harper Collins, London.

Hollom, P.A.D. *et al*. 1988. *Birds of the Middle East and North Africa*. T. & A.D. Poyser, London.

Johnsgard, P.A. 1981. *The plovers, sandpipers, and snipes of the world*. University of Nebraska Press, London and Lincoln.

Jonsson, L. 1992. *Birds of Europe with North Africa and the Middle East*. Christopher Helm, London.

Juniper, J. and Parr, M. 1998. *Parrots: a guide to the parrots of the world*. Pica Press, Surrey.

Kemp, A. and Kemp, M. 1998. *Birds of prey of Africa and its islands*. New Holland, London.

Maclean, G.L. 1993. *Robert's birds of southern Africa*. The Trustees of the John Voelcker Bird Book Fund, Cape Town.

Monroe, B.L. and Sibley, C.G. 1993. *A world checklist of birds*. Yale University Press, New Haven and London.

Mundy, P. *et al*. 1992. *The vultures of Africa*. Acorn Books, Randburg.

Newman, K. 1998. *Newman's birds of southern Africa*. Southern Book Publishers, Halfway House; 7th edition.

Perlo, B. van. 1995. *Birds of Eastern Africa – illustrated checklist*. Harper Collins, London.

Porter, R.F. *et al*. 1996. *Field guide to the birds of the Middle East*. T. & A.D. Poyser, London.

Serle, W. *et al*. 1998. *Birds of West Africa*. Harper Collins, London.

Sinclair, I. 1991. *Field guide to the birds of southern Africa*. Struik Publishers, Cape Town.

Sinclair, I. *et al*. 1997. *Sasol birds of southern Africa*. Struik Publishers, Cape Town.

Steyn, P. 1982. *Birds of prey of southern Africa*. David Philip, Cape Town.

Zimmerman, D. *et al*. 1996. *Birds of Kenya and northern Tanzania*. Russel Friedman Books, Halfway House.

Complete list of the birds of Africa

This list presents, in taxonomic order, all the bird species recorded on the continent of Africa and its associated inshore islands. The following abbreviations are used to indicate birds whose status is cause for concern:

E = endangered

V = vulnerable

R = rare

Order:
STRUTHIONIFORMES

FAMILY: STRUTHIONIDAE (OSTRICH)

Ostrich *Struthio camelus*

Order:
SPHENISCIFORMES

FAMILY: SPHENISCIDAE (PENGUINS)

Penguin, king *Aptenodytes patagonicus*
rockhopper *Eudyptes chrysocome*
macaroni *E. chrysolophus*
jackass *Sphenicus demersus*
Gentoo *Pygoscelis papua*

Order:
GAVIIFORMES

FAMILY: GAVIIDAE (DIVERS)

Diver, red-throated *Gavia stellata*
black-throated *G. arctica*
great northern *G. immer*

Order:
PODICIPEDIFORMES

FAMILY: PODICIPEDIDAE (GREBES)

Grebe, little *Tachybaptus ruficollis*
red-necked *Podiceps grisegena*
great crested *P. cristatus*
black-necked *P. nigricollis*
Slavonian *P. auritus*

Order:
PROCELLARIIFORMES

FAMILY: DIOMEDEIDAE (ALBATROSSES)

Subfamily: Diomedeinae

Albatross, wandering *Diomedea exulans*
snowy *D. chionoptera*
royal *D. epomophora*
Mollymawk, black-browed *D. melanophris*
Tasmanian (shy albatross) *D. cauta*
yellow-nosed *D. chlororhynchos*
grey-headed *D. chrysostoma*
Buller's *D. bulleri*
sooty *Phoebetria fusca*
light-mantled *P. palpebrata*

FAMILY: PROCELLARIIDAE (PETRELS, SHEARWATERS, PRIONS, FULMARS)

Petrel, Antarctic giant *Macronectes giganteus*
Hall's giant *M. halli*
Bulwer's *Bulweria bulwerii*
Jouanin's *B. fallax*
grey *Procellaria cinerea*
white-chinned *P. aequinoctialis*
Antarctic *Thalassoica antarctica*
Cape *Daption capense*
great-winged *Pterodroma macroptera*
white-headed *P. lessonii*
Atlantic *P. incerta*
Kerguelen *P. brevirostris*
herald *P. arminjoniana*
Madeira *P. madeira* **E**
Cape Verde *P. feae* **R**
soft-plumaged *P. mollis*
blue *Halobaena caerulea*
Fulmar, Antarctic *Fulmarus glacialoides*
northern *F. glacialis*
Prion, broad-billed *Pachyptila vittata*
fairy *P. turtur*
slender-billed *P. belcheri*
Antarctic *P. desolata*
fulmar *P. crassirostris*
Shearwater, flesh-footed *Puffinus carneipes*
great *P. gravis*
wedge-tailed *P. pacificus*
sooty *P. griseus*
Manx *P. puffinus*
Mascarene *P. artrodorsalis*
Mediterranean *P. yelkouan*
little *P. assimilis*
Audubon's *P. lherminieri*
Cory's *Calonectris diomedea*
streaked *C. leucomelas*

FAMILY: PELECANOIDIDAE (DIVING PETRELS)

Diving petrel, South Georgia *Pelecanoides georgicus*
common *P. urinatrix*

FAMILY: OCEANITIDAE/HYDROBATIDAE (STORM PETRELS)

Storm petrel, Wilson's *Oceanites oceanicus*
grey-backed *Garrodia nereis*
white-faced *Pelagodroma marina*
black-bellied *Fregetta tropica*
white-bellied *F. grallaria*
European *Hydrobates pelagicus*
band-rumped *Oceanodroma castro*
Swinhoe's *O. monorhis*
Leach's *O. leucorhoa*
Matsudaira's *O. matsudairae*

Order:
PELICANIFORMES

FAMILY: PHAETHONTIDAE (TROPICBIRDS)

Tropicbird, red-billed *Phaethon aethereus*

red-tailed *P. rubricauda*
white-tailed *P. lepturus*

FAMILY: PELECANIDAE (PELICANS)

Pelican, great white *Pelecanus onocrotalus*
pink-backed *P. rufescens*
Dalmatian *P. crispus*

FAMILY: SULIDAE (GANNETS, BOOBIES)

Gannet, northern *Morus bassana*
Cape *M. capensis*
Australian *M. serrator*
Booby, masked *Sula dactylatra*
red-footed *S. sula*
brown *S. leucogaster*

FAMILY: PHALACROCORACIDAE (CORMORANTS)

Cormorant, great *Phalacrocorax carbo*
Cape *P. capensis*
Socotra *P. nigrogularis*
bank *P. neglectus*
long-tailed *P. africanus*
crowned *P. coronatus*
pygmy *P. pygmeus*
Shag, European *P. aristotelis*

FAMILY: ANHINGIDAE (DARTERS)

Darter, African *Anhinga melanogaster*

FAMILY: FREGATIDAE (FRIGATEBIRDS)

Frigatebird, Ascension *Fregata aquila* **R**
magnificent *F. magnificens*
great *F. minor*
lesser *F. ariel*

Order:
CICONIIFORMES

FAMILY: ARDEIDAE (HERONS, EGRETS, BITTERNS)

Heron, grey *Ardea cinerea*
black-headed *A. melanocephala*
goliath *A. goliath*
purple *A. purpurea*
Egret, great *Casmerodius alba*
dimorphic *Egretta dimorpha*
slaty *E. vinaceigula*
Heron, black *E. ardesiaca*
Egret, intermediate *Mesophoyx intermedia*
little *Egretta garzetta*
little blue *E. caerulea*
western reef *E. gularis*
cattle *Bubulcus ibis*
squacco *Ardeola ralloides*
Madagascar pond *A. idae*
rufous-bellied *A. rufiventris*
striated *Butorides striatus*
Night heron, black-crowned *Nycticorax nycticorax*
white-backed *Gorsachius leuconotus*
Bittern, white-crested *Tigriornis leucolophus*
little *Ixobrychus minutus*

dwarf *I. sturmii*
great *Botaurus stellaris*

FAMILY: SCOPIDAE (HAMERKOP)

Hamerkop *Scopus umbretta*

FAMILY: BALAENICIPITIDAE (SHOEBILL)

Shoebill *Balaeniceps rex*

FAMILY: CICONIIDAE (STORKS)

Subfamily: Ciconiinae

Stork, yellow-billed *Mycteria ibis*
Open-bill, African *Anastomus lamelligerus*
Stork, Abdim's *Ciconia abdimii*
    white *C. ciconia*
    woolly-necked *C. episcopus*
    black *C. nigra*
    saddle-billed *Ephippiorhynchus senegalensis*
    marabou *Leptoptilos crumeniferus*

FAMILY: PLATALEIDAE/THRESKIORNITHIDAE (IBISES, SPOONBILLS)

Ibis, sacred *Threskiornis aethiopicus*
Waldrapp *Geronticus eremita* **E**
Ibis, bald *G. calvus* **R**
    wattled *Bostrychia carunculata*
Hadeda *B. hagedash*
Ibis, olive *B. olivacea*
    spot-breasted *B. rara*
    glossy *Plegadis falcinellus*
Spoonbill, African *Platalea alba*
    Eurasian *P. leucorodia*

Order:
PHOENICOPTERIFORMES

FAMILY: PHOENICOPTERIDAE (FLAMINGOS)

Flamingo, lesser *Phoeniconaias minor*
    greater *Phoenicopterus ruber*

Order:
ANSERIFORMES

FAMILY: DENDROCYGNIDAE (WHISTLING DUCKS)

Whistling duck, fulvous *Dendrocygna bicolor*
    white-faced *D. viduata*

FAMILY: ANATIDAE (DUCKS, GEESE, SWANS)

Swan, mute *Cygnus olor*
    whooper *C. cygnus*
    Bewick's *C. columbianus*
Goose, greylag *Anser anser*
    greater white-fronted *A. albifrons*
    bean *A. fabalis*
    lesser white-fronted *A. erythropus*
    snow *A. caerulescens*
    blue-winged *Cyanochen cyanopterus*
    Egyptian *Alopochen aegyptiacus*
    barnacle *Branta leucopsis*
    Brent *B. bernicla*
    red-breasted *B. ruficollis*
Shelduck, South African *Tadorna cana*

ruddy *T. ferruginea*
    common *T. tadorna*
Duck, African black *Anas sparsa*
Mallard *A. platyrhynchos*
Duck, yellow-billed *A. undulata*
Gadwall *A. strepera*
Teal, common *A. crecca*
Wigeon, Eurasian *A. penelope*
Pintail, northern *A. acuta*
Teal, red-billed *A. erythrorhyncha*
    Cape *A. capensis*
    Hottentot *A. hottentota*
Garganey *A. querquedula*
Shoveler, Cape *A. smithii*
    northern *A. clypeata*
Wigeon, American *A. americana*
Teal, blue-winged *A. discors*
    marbled *Marmaronetta angustirostris*
Eider, common *Somateria mollissima*
Pochard, red-crested *Netta rufina*
    southern *N. erythrophthalma*
    common *Aythya ferina*
Duck, ferruginous *A. nyroca*
    tufted *A. fuligula*
Scaup, great *A. marila*
Duck, ring-necked *A. collaris*
Goose, African pygmy *Nettapus auritus*
Duck, white-backed *Thalassornis leuconotus*
    comb *Sarkidiornis melanotos*
    Hartlaub's *Pteronetta hartlaubi*
Goose, spurwinged *Plectropterus gambensis*
Scoter, black *Melanitta nigra*
    velvet *M. fusca*
Goldeneye, common *Bucephala clangula*
Smew *Mergus albellus*
Merganser, red-breasted *M. serrator*
    common *M. merganser*
Duck, white-headed *Oxyura leucocephala*
    Maccoa *O. maccoa*
    mandarin *Aix galericulata*

Order:
FALCONIFORMES

FAMILY: SAGITTARIIDAE (SECRETARYBIRD)

Secretarybird *Sagittarius serpentarius*

FAMILY: ACCIPITRIDAE (EAGLES, HAWKS, BUZZARDS, KITES, VULTURES)

Griffon, Eurasian *Gyps fulvus*
    Cape *G. coprotheres* **R**
    Rüppell's *G. rueppelli*
Vulture, white-backed *G. africanus*
    bearded *Gypaetus barbatus*
    palm-nut *Gypohierax angolensis*
    cinereous *Aegypius monachus*
    hooded *Necrosyrtes monachus*
    Egyptian *Neophron percnopterus*
    lappet-faced *Torgos tracheliotus*
    white-headed *Trigonoceps occipitalis*
Eagle, white-tailed *Haliaeetus albicilla*
Fish eagle, African *H. vocifer*
Snake eagle, brown *Circaetus cinereus*
    banded *C. cinerascens*
    fasciated *C. fasciolatus*

short-toed *C. gallicus*
Bateleur *Terathopius ecaudatus*
Eagle, long-crested *Lophaetus occipitalis*
    tawny *Aquila rapax*
    steppe *A. nipalensis*
    lesser spotted *A. pomarina*
    Verreaux's *A. verreauxii*
    Wahlberg's *A. wahlbergi*
    greater spotted *A. clanga*
Hawk-eagle, Ayres's *Hieraaetus ayresii*
Eagle, imperial *Aquila heliaca*
    golden *A. chrysaetos*
Hawk-eagle, crowned *Stephanoaetus coronatus*
Eagle, martial *Polemaetus bellicosus*
    booted *Hieraaetus pennatus*
    Bonelli's *H. fasciatus*
Hawk-eagle, African *H. spilogaster*
    Cassin's *Spizaetus africanus*
Serpent-eagle, Congo *Dryotriorchis spectabilis*
Sparrowhawk, Levant *Accipiter brevipes*
    Eurasian *A. nisus*
    chestnut-flanked *A. castanilius*
    Ovambo *A. ovampensis*
    red-breasted *A. rufiventris*
    little *A. minullus*
    red-thighed *A. erythropus*
Goshawk, northern *A. gentilis*
    African *A. tachiro*
    black *A. melanoleucus*
Shikra *A. badius*
Goshawk, red-chested *A. toussenelii*
    Frances's *A. francesii*
Hawk, long-tailed *Urotriorchis macrourus*
Harrier-hawk, African *Polyboroides typus*
Buzzard, red-necked *Buteo auguralis*
    long-legged *B. rufinus*
    jackal *B. rufofuscus*
    augur *B. augur*
    Archer's *B. archeri*
    common *B. buteo*
    forest *B. tachardus*
    mountain *B. oreophilus*
Hawk, rough-legged *B. lagopus*
Honey buzzard, European *Pernis apivorus*
Buzzard, grasshopper *Butastur rufipennis*
    lizard *Kaupifalco monogrammicus*
Harrier, African marsh *Circus ranivorus*
    black *C. maurus*
    pallid *C. macrourus*
    European marsh *C. aeruginosus*
    Montagu's *C. pygargus*
    hen *C. cyaneus*
Hawk, bat *Macheirhamphus alcinus*
Chanting goshawk, pale *Melierax canorus*
    dark *M. metabates*
    eastern *M. poliopterus*
Goshawk, gabar *Micronisus gabar*
Kite, red *Milvus milvus*
    black *M. migrans*
    yellow-billed *M. migrans parasitus*
    black-shouldered *Elanus caeruleus*
    swallow-tailed *Chelictinia riocourii*
Osprey *Pandion haliaetus*

FAMILY: FALCONIDAE (FALCONS, KESTRELS)

Falcon, peregrine *Falco peregrinus*

lanner *F. biarmicus*
Hobby, African *F. cuvieri*
Eurasian *F. subbuteo*
Falcon, taita *F. fasciinucha*
Eleonora's *F. eleonorae*
red-necked *F. chicquera*
red-footed *F. vespertinus*
Amur *F. amurensis*
sooty *F. concolor*
Saker *F. cherrug*
Merlin *F. columbarius*
Falcon, Barbary *F. pelegrinoides*
Kestrel, Dickinson's *F. dickinsoni*
lesser *F. naumanni*
grey *F. ardosiaceus*
common *F. tinnunculus*
greater *F. rupicoloides*
fox *F. alopex*
Baza, African *Aviceda cuculoides*
Falcon, pygmy *Polihierax semitorquatus*

Order:
GALLIFORMES

FAMILY: PHASIANIDAE (FRANCOLINS, QUAILS)

Peacock, African *Afropavo congensis*
Francolin, Nahan's *Francolinus nahani* **R**
coqui *F. coqui*
Schlegel's *F. schlegelii*
crested *F. sephaena*
ring-necked *F. streptophorus*
white throated *F. albogularis*
Shelley's *F. shelleyi*
Finsch's *F. finschi*
red-winged *F. levaillantii*
double-spurred *F. bicalcaratus*
Heuglin's *F. icterorhynchus*
Clapperton's *F. clappertoni*
Hildebrandt's *F. hildebrandti*
scaly *F. squamatus*
Ahanta *F. ahantensis*
Cameroon *F. camerunensis* **R**
handsome *F. nobilis*
Latham's forest *F. lathami*
moorland *F. psilolaemus*
Jackson's *F. jacksoni*
chestnut-naped *F. castaneicollis*
red-billed *F. adspersus*
Cape *F. capensis*
Natal *F. natalensis*
Orange River *F. levaillantoides*
grey-winged *F. africanus*
Hartlaub's *F. hartlaubi*
grey-striped *F. griseostriatus*
Harwood's *F. harwoodi*
Swierstra's *F. swierstrai*
ochre-breasted *F. ochropectus* **E**
Erckel's *F. ercklii*
Spurfowl, red-necked *F. afer*
grey-breasted *F. rufopictus*
Swainson's *F. swainsonii*
yellow-necked *F. leucoscepus*
Chukar *Alectoris chukar*
Partridge, Barbary *A. barbara*
rock *A. graeca*

sand *Ammoperdix heyi*
Udzungwa *Xenoperdix udzungwensis*
stone *Ptilopachus petrosus*
Quail *Coturnix coturnix*
harlequin *C. delegorguei*
blue *C. chinensis adansoni*
Plover, quail *Ortyxelos meiffrenii*

FAMILY: NUMIDIDAE (GUINEAFOWL)

Guineafowl, helmeted *Numida meleagris*
crested *Guttera pucherani*
plumed *G. plumifera*
vulturine *Acryllium vulturinum*
black *Agelastes niger*
white-breasted *A. meleagrides* **E**

Order:
GRUIFORMES

FAMILY: TURNICIDAE
(BUTTONQUAILS)

Buttonquail, Kurrichane (small)
Turnix sylvatica
Hottentot *T. hottentotta*
black-rumped *T. nana*

FAMILY: GRUIDAE (CRANES)

Crane, wattled *Grus carunculatus*
common *G. grus*
blue *Anthropoides paradiseus*
demoiselle *A. virgo*
grey crowned *Balearica regulorum*
black crowned *B. pavonina*

FAMILY: RALLIDAE (RAILS, CRAKES,
MOORHENS, COOTS, ETC.)

Coot *Fulica atra*
red-knobbed *F. cristata*
Gallinule, purple (swamp hen) *Porphyrio
porphyrio*
lesser *P. alleni*
American purple *P. martinicus*
Moorhen, common *Gallinula chloropus*
lesser *G. angulata*
Crake, black *Amaurornis flavirostra*
striped *Aenigmatolimnas marginalis*
Rail, Rouget's *Rallus rougetii*
African *R. caerulescens*
water *R. aquaticus*
grey-throated *Canirallus oculeus*
Nkulengu *Himantornis haematopus*
Corncrake *Crex crex*
Crake, African *C. egregia*
spotted *Porzana porzana*
Baillon's *P. pusilla*
little *P. parva*
Flufftail, white-spotted *Sarothrura pulchra*
Crake, white-winged *S. ayresi*
chestnut-tailed *S. lineata*
Lyne's *S. lynesi*
Flufftail, red-chested *S. rufa*
buff-spotted *S. elegans*
streaky-breasted *S. boehmi*
chestnut-headed *S. lugens*
striped *S. affinis*

FAMILY: HELIORNITHIDAE (FINFOOTS)

Finfoot, African *Podica senegalensis*

FAMILY: OTIDIDAE (BUSTARDS AND KORHAANS)

Bustard, Stanley's *Neotis denhami*
Ludwig's *N. ludwigii*
Nubian *N. nuba*
Heuglin's *N. heuglinii*
great *Otis tarda*
kori *Ardeotis kori*
Arabian *A. arabs*
Rüppell's *Eupodotis rueppellii*
Karoo *E. vigorsii*
red-crested *E. ruficrista*
black-bellied *E. melanogaster*
white-bellied *E. senegalensis*
buff-crested *E. gindiana*
blue *E. caerulescens*
black *E. afra*
white-winged *E. afraoides*
Hartlaub's *E. hartlaubii*
Savile's *E. savilei*
houbara *Chlamydotis undulata*
little *Tetrax tetrax*
little brown *Eupodotis humilis*

Order:
CHARADRIIFORMES

FAMILY: JACANIDAE (JACANAS)

Jacana, African *Actophilornis africanus*
lesser *Microparra capensis*

FAMILY: ROSTRATULIDAE (PAINTED SNIPE)

Snipe, painted *Rostratula benghalensis*

FAMILY: HAEMATOPODIDAE (OYSTERCATCHERS)

Oystercatcher, black *Haematopus moquini*
European (Eurasian) *H. ostralegus*

FAMILY: CHARADRIIDAE (PLOVERS)

Lapwing, black-winged *Vanellus melanopterus*
blacksmith *V. armatus*
crowned *V. coronatus*
long-toed *V. crassirostris*
white-tailed *V. leucurus*
spur-winged *V. spinosus*
white-headed *V. albiceps*
Senegal *V. lugubris*
spot-breasted *V. melanocephalus*
wattled *V. senegallus*
brown-chested *V. superciliosus*
black-headed *V. tectus*
northern *V. vanellus*
sociable *V. gregarius*
Plover, Kentish *Charadrius alexandrinus*
Caspian *C. asiaticus*
little ringed *C. dubius*
Forbe's *C. forbesi*
common ringed *C. hiaticula*
great sand *C. leschenaultii*
white-fronted *C. marginatus*
Mongolian *C. mongolus*
Kittlitz's *C. pecuarius*

three-banded *C. tricollaris*
chestnut-banded *C. pallidus*
Dotterel, Eurasian *Eudromias morinellus*
Golden plover, Eurasian *Pluvialis apricaria*
American *P. dominica*
Pacific *P. fulva*
Plover, grey *P. squatarola*

FAMILY: SCOLOPACIDAE (TURNSTONES, SANDPIPERS, STINTS, SNIPES, CURLEWS, PHALAROPES, ETC.)

Dowitcher, Asian *Limnodromus semipalmatus*
long-billed *L. scolopaceus*
Curlew, Eurasian *Numenius arquata*
slender-billed *N. tenuirostris*
Whimbrel *N. phaeopus*
Godwit, bar-tailed *Limosa lapponica*
black-tailed *L. limosa*
Hudsonian *L. haemastica*
Sandpiper, terek *Tringa cinerea*
Redshank, spotted *T. erythropus*
Sandpiper, wood *T. glareola*
Greenshank, common *T. nebularia*
Sandpiper, green *T. ochropus*
marsh *T. stagnatilis*
solitary *T. solitaria*
Yellowlegs, lesser *T. flavipes*
greater *T. melanoleuca*
Redshank, common *T. totanus*
Sandpiper, buff-breasted *Tryngites subruficollis*
common *T. hypoleucos*
Snipe, common *Gallinago gallinago*
great *G. media*
African *G. nigripennis*
pintail *G. stenura*
Jack *Lymnocryptes minimus*
Turnstone, ruddy *Arenaria interpres*
Stint, rufous-necked *Calidris ruficollis*
Sanderling *C. alba*
Dunlin *C. alpina*
Sandpiper, curlew *C. ferruginea*
Stint, little *C. minuta*
Sandpiper, Baird's *C. bairdii*
white-rumped *C. fuscicollis*
pectoral *C. melanotos*
Stint, Temminck's *C. temminckii*
long-toed *C. subminuta*
Sandpiper, purple *C. maritima*
Knot, great *C. tenuirostris*
red *C. canutus*
Woodcock, Eurasian *Scolopax rusticola*
Sandpiper, broad-billed *Limicola falcinellus*
Ruff *Philomachus pugnax*
Phalarope, red *Phalaropus fulicarius*
red-necked *P. lobatus*
Wilson's *Steganopus tricolor*

FAMILY: RECURVIROSTRIDAE (AVOCETS, STILTS)

Avocet *Recurvirostra avosetta*
Stilt, black-winged *Himantopus himantopus*

FAMILY: DROMADIDAE (CRAB PLOVER)

Plover, crab *Dromas ardeola*

FAMILY: BURHINIDAE (DIKKOPS/THICK-KNEES)

Dikkop, Cape (Thick-knee, spotted)

Burhinus capensis
Thick-knee, Senegal *B. senegalensis*
Dikkop (Thick-knee), water *B. vermiculatus*
Curlew, stone *B. oedicnemus*

FAMILY: GLAREOLIDAE (COURSERS, PRATINCOLES)

Bird, crocodile *Pluvianus aegyptius*
Pratincole, black-winged *Glareola nordmanni*
collared *G. pratincola*
rock *G. nuchalis*
Madagascar *G. ocularis*
grey *Galachrysia cinerea*
Courser, Temminck's *Cursorius temminckii*
Burchell's *C. rufus*
cream-coloured *C. cursor*
double-banded *Rhinoptilus africanus*
three-banded *R. cinctus*
bronze-winged *R. chalcopterus*

FAMILY: CHIONIDIDAE (SHEATHBILLS)

Sheathbill, greater *Chionis alba*

FAMILY: LARIDAE (SKUAS, GULLS, TERNS)

Gull, white-eyed *Larus leucophthalmus*
sooty *L. hemprichi*
Audouin's *L. audouinii*
common *L. canus*
ring-billed *L. delawarensis*
yellow-legged *L. cachinnans*
herring *L. argentatus*
lesser black-backed *L. fuscus*
great black-backed *L. marinus*
great black-headed *L. ichthyaetus*
grey-headed *L. cirrocephalus*
Franklin's *L. pipixcan*
Hartlaub's *L. hartlaubii*
Mediterranean *L. melanocephalus*
common black-headed *L. ridibundus*
slender-billed *L. genei*
little *L. minutus*
Sabine's *Xema sabini*
laughing *Larus atricilla*
kelp *L. dominicanus*
glaucous *L. hyperboreus*
Kittiwake, black-legged *Rissa tridactyla*
Skua, great *Catharacta skua*
South Polar *C. maccormicki*
Antarctic *C. antarctica*
Jaeger, pomarine *Stercorarius pomarinus*
parasitic *S. parasiticus*
long-tailed *S. longicaudus*
Tern, whiskered *Chlidonias hybridus*
white-winged *C. leucopterus*
black *C. niger*
gull-billed *Sterna nilotica*
Caspian *S. caspia*
common *S. hirundo*
Arctic *S. paradisaea*
Antarctic *S. vittata*
roseate *S. dougallii*
white-cheeked *S. repressa*
black-naped *S. sumatrana*
bridled *S. anaethetus*
sooty *S. fuscata*
Damara *S. balaenarum* **R**

little *S. albifrons*
royal *S. maxima*
lesser crested *S. bengalensis*
sandwich *S. sandvicensis*
great crested *S. bergii*
Saunders's *S. saundersi*
Noddy, brown *Anous stolidus*
lesser *A. tenuirostris*
black *A. minutus*
Tern, white (fairy) *Gygis alba*

FAMILY: RYNCHOPIDAE (SKIMMERS)

Skimmer, African *Rynchops flavirostris*

FAMILY: ALCIDAE (AUKS)

Auk, little *Alle alle*
Razorbill *Alca torda*
Guillemot *Uria aalge*
Puffin, Atlantic *Fratercula arctica*

Order:
PTEROCLIFORMES

FAMILY: PTEROCLIDAE (SANDGROUSE)

Sandgrouse, double-banded *Pterocles bicinctus*
yellow-throated *P. gutturalis*
Namaqua *P. namaqua*
Burchell's *P. burchellii*
spotted *P. senegallus*
chestnut-bellied *P. exustus*
black-faced *P. decoratus*
black-bellied *P. orientalis*
pin-tailed *P. alchata*
crowned *P. coronatus*
Lichtenstein's *P. lichtensteinii*
four-banded *P. quadricinctus*

Order:
COLUMBIFORMES

FAMILY: COLUMBIDAE (PIGEONS, DOVES)

Dove, rock *Columba livia*
stock *C. oenas*
Wood pigeon, common *C. palumbus*
Pigeon, speckled *C. guinea*
Somali *C. oliviae* **R**
white-naped *C. albinucha*
white-collared *C. albitorques*
afep *C. unicincta*
eastern bronze-necked *C. delegorguei*
African olive *C. arquatrix*
São Tomé olive *C. thomensis*
Cameroon *C. sjostedti*
Comoro *C. pollenii*
Trocaz *C. trocaz* **R**
Bolle's *C. bollii* **R**
laurel *C. junoniai* **R**
Dove, lemon *C. larvata*
African collared *Streptopelia roseogrisea*
Eurasian *S. dacaocto*
red-eyed *S. semitorquata*
European turtle *S. turtur*
palm *S. senegalensis*
dusky turtle *S. lugens*
mourning collared *S. decipiens*

Dove, ring-necked *S. capicola*
  African white-winged collared *S. reichenowi*
  vinaceous *S. vinacea*
  Adamawa turtle *S. hypopyrrha*
  Namaqua *Oena capensis*
  tambourine *Turtur tympanistria*
Wood dove, emerald-spotted *T. chalcospilos*
  blue-spotted *T. afer*
  black-billed *T. abyssinicus*
  blue-headed *T. brehmeri*
Pigeon, Gabon bronze-naped *Columba iriditorques*
  São Tomé bronze-naped *C. malherbii*
  Comoro blue *Alectroenas sganzini*
Green pigeon, African *Treron calva*
  Bruce's *T. waalia*
  São Tomé *T. sanctithomae*
  Pemba *T. pembaensis*

Order: PSITTACIFORMES

FAMILY: PSITTACIDAE (PARROTS, PARAKEETS, LOVEBIRDS)

Parrot, Meyer's *Poicephalus meyeri*
  Rüppell's *P. rueppellii*
  brown-necked *P. robustus*
  brown-headed *P. cryptoxanthus*
  niam-niam *P. crassus*
  yellow-bellied *P. senegalus*
  red-bellied *P. rufiventris*
  red-fronted *P. gulielmi*
  yellow-fronted *P. flavifrons*
  grey *Psittacus erithacus*
  vasa *Coracopsis vasa*
  black *C. nigra*
Parakeet, rose-ringed *Psittacula krameri*
Lovebird, rosy-faced *Agapornis roseicollis*
  black-cheeked *A. nigrigenis* **R**
  Lilian's *A. lilianae*
  red-headed *A. pullarius*
  black-collared *A. swindernianus*
  yellow-collared *A. personatus*
  grey-headed *A. cana*
  Fischer's *A. fischeri*
  black-winged *A. taranta*

Order: MUSOPHAGIFORMES

FAMILY: MUSOPHAGIDAE (TURACOS, GO-AWAY BIRDS, PLANTAIN-EATERS)

Turaco, Guinea *Tauraco persa*
  Knysna *T. corythaix*
  Schalow's *T. schalowi*
  Livingstone's *Tauraco livingstoni*
  Fischer's *T. fischeri*
  black-billed *T. schuettii*
  Hartlaub's *T. hartlaubi*
  white-crested *T. leucolophus*
  white-cheeked *T. leucotis*
  Prince Ruspoli's *T. ruspolii* **R**
  yellow-billed *T. macrorhynchus*
  red-crested *T. erythrolophus*

Bannerman's *T. bannermani* **R**
  purple-crested *Musophaga porphyreolopha*
  Ross's *M. rossae*
  Ruwenzori *M. johnstoni*
  violet *M. violacea*
  grey *Corythaixoides concolor*
  white-bellied *C. leucogaster*
  bare-faced *C. personatus*
  great blue *Corythaeola cristata*
Plantain-eater, eastern grey *Crinifer zonurus*
  western grey *C. piscator*

Order: CUCULIFORMES

FAMILY: CUCULIDAE (CUCKOOS)

Cuckoo, Eurasian *Cuculus canorus*
  African *C. gularis*
  lesser *C. poliocephalus*
  red-chested *C. solitarius*
  black *C. clamosus*
  Madagascar *C. rochii*
  black & white *Oxylophus jacobinus*
  Levaillant's *O. levaillantii*
  great spotted *Clamator glandarius*
  thick-billed *Pachycoccyx audeberti*
  African emerald *Chrysococcyx cupreus*
  Klaas's *C. klaas*
  Diederik *C. caprius*
  yellow-throated *C. flavigularis*
  dusky long-tailed *Cercococcyx mechowi*
  olive long-tailed *C. olivinus*
  barred long-tailed *C. montanus*
Yellowbill *Ceuthmochares aereus*

FAMILY: CENTROPODIDAE (COUCALS)

Coucal, coppery-tailed *Centropus cupreicaudus*
  white-browed *C. superciliosus*
  Burchell's *C. burchelli*
  Senegal *C. senegalensis*
  Neumann's *C. neumanni*
  blue-headed *C. monachus*
  black *C. grillii*
  black-throated *C. leucogaster*
  Gabon *C. anselli*

Order: STRIGIFORMES

FAMILY: TYTONIDAE (BARN OWLS, GRASS OWLS, BAY OWLS)

Owl, barn *Tyto alba*
  African grass *T. capensis*
Congo bay *Phodilus prigoginei*

FAMILY: STRIGIDAE (TYPICAL OWLS)

Owl, Comoro scops *Otus pauliani* **R**
  sandy scops *O. icterorhynchus*
  Bruce's scops *O. brucei*
  common scops *O. scops*
  Pemba scops *O. rutilus*
  Sokoke scops *O. ireneae* **E**
  São Tomé scops *O. hartlaubi* **R**
  white-faced scops *O. leucotis*

Eagle owl, Usambara *Bubo vosseleri* **R**
  Shelley's *B. shelleyi*
  Eurasian *B. bubo*
  Pharao *B. ascalaphus*
  spotted *B. africanus*
  Verraux's *B. lacteus*
  Mackinder's (Cape) *B. capensis*
  Fraser's *B. poensis*
  Akun *B. leucostictus*
Owl, wood *Ciccaba woodfordi*
  little *Athene noctua*
  tawny *Strix aluco*
  Hume's tawny *S. butleri*
  long-eared *Asio otus*
  Abyssinian long-eared *A. abyssinicus*
  short-eared *A. flammeus*
  marsh *A. capensis*
  Pel's fishing *Scotopelia peli*
  rufous fishing *S. ussheri* **R**
  vermiculated fishing *S. bouvieri*
  pearl-spotted *Glaucidium perlatum*
Owlet, Sjostedt's *G. sjostedti*
  red-chested *G. tephronotum*
  African barred *G. capense*
  chestnut *G. castaneum*
  Ngami *G. ngamiense*
  Scheffler's *G. scheffleri*
  Albertine *G. albertinum* **R**
Owl, maned *Jubula lettii*

Order: CAPRIMULGIFORMES

FAMILY: CAPRIMULGIDAE (NIGHTJARS)

Nightjar, plain *Caprimulgus inornatus*
  Nubian *C. nubicus*
  European *C. europaeus*
  red-necked *C. ruficollis*
  Egyptian *C. aegyptius*
  fiery-necked *C. pectoralis*
  montane *C. poliocephalus*
  sombre *C. fraenatus*
  swamp *C. natalensis*
  star-spotted *C. stellatus*
  Donaldson-Smith's *C. donaldsoni*
  freckled *C. tristigma*
  Bates *C. batesi*
  slender-tailed *C. clarus*
  Nechisar *C. solala*
  long-tailed *C. climacurus*
  square-tailed *C. fossii*
  rufous-cheeked *C. rufigena*
  Ruwenzori *C. ruwenzorii*
  golden *C. eximius*
  brown *C. binotatus*
  black-shouldered *C. nigriscapularis*
  Itombwe *C. prigoginei*
  standard-winged *Macrodipteryx longipennis*
  pennant-winged *M. vexillaria*

Order: APODIFORMES

FAMILY: APODIDAE (SWIFTS, SPINETAILS)

Swift, pallid *Apus pallidus*
  European *A. apus*
  Alexander's *A. alexandri*
  African *A. barbatus*
  Bradfield's *A. bradfieldi*
  white-rumped *A. caffer*
  horus *A. horus*
  little *A. affinis*
  plain *A. unicolor*
  Loanda *A. toulsoni*
  Bates's *A. batesi*
  Nyanza *A. niansae*
  Forbes-Watson's *A. berliozi*
  scarce *Schoutedenapus myoptilus*
  Schouteden's *S. schoutedeni*
  alpine *Tachymarptis melba*
  mottled *T. aequatorialis*
  African palm *Cypsiurus parvus*
Spinetail, bat-like *Neafrapus boehmi*
  Cassin's *N. cassini*
  mottled *Telacanthura ussheri*
  black *T. melanopygia*
  Sabine's *Raphidura sabini*
  São Tomé *Zoonavena thomensis*
  Malagasy *Z. grandidieri*

Order:
COLIIFORMES

FAMILY: COLIIDAE (MOUSEBIRDS)

Mousebird, speckled *Colius striatus*
  white-backed *C. colius*
  white-headed *C. leucocephalus*
  red-backed *C. castanotus*
  red-faced *Urocolius indicus*
  blue-naped *U. macrourus*

Order:
TROGONIFORMES

FAMILY: TROGONIDAE (TROGONS)

Trogon, Narina *Apaloderma narina*
  bar-tailed *A. vittatum*
  bare-cheeked *A. aequatoriale*

Order:
CORACIIFORMES

FAMILY: ALCEDINIDAE/HALCYONIDAE (KINGFISHERS)

Kingfisher, half-collared *Alcedo semitorquata*
  shining blue *A. quadribrachys*
  common *A. atthis*
  malachite *A. cristata*
  Principe *A. nais*
  São Tomé *A. thomensis*
  Madagascar *A. vintsioides*
  giant *Ceryle maxima*
  pied *C. rudis*
  African pygmy *Ispidina picta*
  African dwarf *I. lecontei*
  woodland *Halcyon senegalensis*
  mangrove *H. senegaloides*
  brown-hooded *H. albiventris*
  grey-headed *H. leucocephala*

  striped *H. chelicuti*
  chocolate-backed *H. badia*
  blue-breasted *H. malimbica*
  white-collared *Todirhamphus chloris*
  white-breasted *Halcyon smyrnensis*
  white-bellied *Alcedo leucogaster*

FAMILY: MEROPIDAE (BEE-EATERS)

Bee-eater, blue-headed *Merops muelleri*
  black *M. gularis*
  little *M. pusillus*
  blue-breasted *M. variegatus*
  cinnamon-chested *M. oreobates*
  swallow-tailed *M. hirundineus*
  northern carmine *M. nubicus*
  southern carmine *M. nubicoides*
  white-fronted *M. bullockoides*
  red-throated *M. bullocki*
  Somali *M. revoilii*
  white-throated *M. albicollis*
  little green *M. orientalis*
  Böhm's *M. boehmi*
  blue-cheeked *M. persicus*
  olive *M. superciliosus*
  European *M. apiaster*
  black-headed *M. breweri*
  rosy *M. malimbicus*

FAMILY: CORACIIDAE (ROLLERS)

Roller, racket-tailed *Coracias spatulata*
  rufous-crowned *C. naevia*
  European *C. garrulus*
  lilac-breasted *C. caudata*
  Abyssinian *C. abyssinica*
  blue-bellied *C. cyanogaster*
  broad-billed *Eurystomus glaucurus*
  blue-throated *E. gularis*

Order:
UPUPIFORMES

FAMILY: UPUPIDAE (HOOPOES)

Hoopoe, Eurasian *Upupa epops*
  African *U. africana*

FAMILY: PHOENICULIDAE (WOOD-HOOPOES)

Wood-hoopoe, violet *Phoeniculus damarensis*
  green *P. purpureus*
  white-headed *P. bollei*
  forest *P. castaneiceps*
  black-billed *P. somaliensis*
Scimitar bill, common *Rhinopomastus cyanomelas*
  black *R. aterrimus*
  Abyssinian *R. minor*

Order:
BUCEROTIFORMES

FAMILY: BUCEROTIDAE (HORNBILLS)

Hornbill, African grey *Tockus nasutus*
  pale-billed *T. pallidirostris*
  red-billed *T. erythrorhychus*
  von der Decken's *T. deckeni*
  Jackson's *T. jacksoni*

  southern yellow-billed *T. leucomelas*
  eastern yellow-billed *T. flavirostris*
  Hemprich's *T. hemprichii*
  crowned *T. alboterminatus*
  African pied *T. fasciatus*
  red-billed dwarf *T. camurus*
  black dwarf *T. hartlaubi*
  white-crested *T. albocristatus*
  Bradfield's *T. bradfieldi*
  Monteiro's *T. monteiri*
  black & white casqued *Ceratogymna subcylindricus*
  piping *C. fistulator*
  yellow-casqued *C. elata*
  black-casqued *C. atrata*
  trumpeter *C. bucinator*
  silvery-cheeked *C. brevis*
  white-thighed *C. albotibialis*
  brown-cheeked *C. cylindricus*

FAMILY: BUCORVIDAE (GROUND HORNBILLS)

Hornbill, southern ground *Bucorvus leadbeateri*
  Abyssinian ground *B. abyssinicus*

Order:
PICIFORMES

FAMILY: CAPITONIDAE (BARBETS, TINKERBIRDS)

Barbet, banded *Lybius undatus*
  Vieillot's *L. vieilloti*
  white-headed *L. leucocephalus*
  red-faced *L. rubrifacies*
  black-billed *L. guifsobalito*
  black-collared *L. torquatus*
  brown-breasted *L. melanopterus*
  black-backed *L. minor*
  double-toothed *L. bidentatus*
  black-breasted *L. rolleti*
  bearded *L. dubius*
  Chaplin's *L. chaplini*
  pied *Tricholaema leucomelas*
  hairy-breasted *T. hirsuta*
  red-fronted *T. diademata*
  miombo pied *T. frontata*
  spot-flanked *T. lachrymosa*
  black-throated *T. melanocephala*
  Woodward's *Cryptolybia woodwardi*
  white-eared *Stactolaema leucotis*
  Whyte's *S. whytii*
  green *S. olivacea*
  Anchieta's *S. anchietae*
  yellow-spotted *Buccanodon duchaillui*
  grey-throated *Gymnobucco bonapartei*
  bristle-nosed *G. peli*
  Sladen's *G. sladeni*
  naked-faced *G. calvus*
  yellow-billed *Trachyphonus purpuratus*
  yellow-breasted *T. margaritatus*
  D'Arnaud's *T. darnaudii*
  Usambiro *T. usambiro*
  crested *T. vaillantii*
  red & yellow *T. erythrocephalus*
Tinkerbird, speckled *Pogoniulus scolopaceus*
  moustached *P. leucomystax*

red-rumped *P. atroflavus*
green *P. simplex*
western *P. coryphaeus*
yellow-throated *P. subsulphureus*
red-fronted *P. pusillus*
yellow-rumped *P. bilineatus*
yellow-fronted *P. chrysoconus*
white-chested *P. makawai*

FAMILY: INDICATORIDAE
(HONEYGUIDES)

Honeyguide, pallid *Indicator meliphilus*
lesser *I. minor*
scaly-throated *I. variegatus*
greater *I. indicator*
spotted *I. maculatus*
thick-billed *I. conirostris*
least *I. exilis*
Willcock's *I. willcocksi*
dwarf *I. pumilio*
green-backed *Prodotiscus zambesiae*
Wahlberg's *P. regulus*
Cassin's *P. insignis*
Zenker's *Melignomon zenkeri*
yellow-footed *M. eisentrauti*
lyre-tailed *Melichneutes robustus*

FAMILY: PICIDAE (WOODPECKERS,
PICULETS, WRYNECKS)

Woodpecker, green *Picus viridis*
Levaillant's *P. vaillantii*
great spotted *Dendrocopos major*
lesser spotted *D. minor*
ground *Geocolaptes olivaceus*
Bennett's *Campethera bennettii*
Reichenow's *C. scriptoricauda*
Knysna *C. notata*
golden-tailed *C. abingoni*
little green *C. maculosa*
green-backed *C. cailliautii*
brown-eared *C. caroli*
buff-spotted *C. nivosa*
Tullberg's *C. tullbergi*
Mombasa *C. mombassica*
Nubian *C. nubica*
fine-spotted *C. punctuligera*
brown-backed *Dendropicos obsoletus*
cardinal *D. fuscescens*
little grey *D. elachus*
speckle-breasted *D. poecilolaemus*
Gabon *D. gabonensis*
melancholy *D. lugubris*
olive *D. griseocephalus*
bearded *D. namaquus*
grey *D. goertae*
fire-bellied *D. pyrrhogaster*
grey-headed *D. spodocephalus*
golden-crowned *D. xantholophus*
Elliot's *D. elliotii*
Abyssinian *D. abyssinicus*
Stierling's *D. stierlingi*
Piculet, African *Sasia africana*
Wryneck, Eurasian *Jynx torquilla*
rufous-necked *J. ruficollis*

Order:
PASSERIFORMES

FAMILY: EURYLAIMIDAE (BROADBILLS)

Broadbill, African *Smithornis capensis*
rufous-sided *S. rufolateralis*
grey-headed *S. sharpei*
Grauer's *Pseudocalyptomena graueri* **R**

FAMILY: PITTIDAE (PITTAS)

Pitta, African *Pitta angolensis*
green-breasted *P. reichenowi*

FAMILY: ALAUDIDAE (LARKS)

Lark, flappet *Mirafra rufocinnamomea*
monotonous *M. passerina*
Somali *M. sharpii*
clapper *M. apiata*
rufous-naped *M. africana*
Abyssinian *M. alopex*
sabota *M. sabota*
melodious *M. cheniana*
Bradfield's *M. naevia*
fawn-coloured *M. africanoides*
Degodi *M. degodiensis*
African singing bush *M. cantillans*
white-tailed *M. albicauda*
collared *M. collaris*
red-winged *M. hypermetra*
Somali long-billed *M. somalica*
Angola *M. angolensis*
pink-breasted *M. poecilosterna*
Gillett's *M. gilletti*
William's *M. williamsi*
Friedmann's *M. pulpa*
Ash's *M. ashi*
kordofan *M. cordofanica*
rusty *M. rufa*
dusky *Pinarocorys nigricans*
rufous-rumped *P. erythropygia*
thick-billed *Galerida magnirostris*
sun *G. modesta*
crested *G. cristata*
thekla *G. theklae*
bimaculated *Melanocorypha bimaculata*
calandra *M. calandra*
long-billed *Certhilauda curvirostris*
short-clawed *C. chuana*
Karoo *C. albescens*
red *C. burra*
dune *C. erythrochlamys*
Barlow's *C. barlowi*
Rudd's *Heteromirafra ruddi*
sidamo *H. sidamoensis*
Archer's *H. archeri*
spike-heeled *Chersomanes albofasciata*
red-capped *Calandrella cinerea*
Blanford's *C. blanfordi*
greater short-toed *C. brachydactyla*
Erlanger's *C. erlangeri*
lesser short-toed *C. rufescens*
rufous short-toed *C. somalica*
Athi short-toed *C. athensis*
Dunn's *C. dunni*
Botha's *Spizocorys fringillaris*

pink-billed *S. conirostris*
Sclater's *S. sclateri*
Stark's *Eremalauda starki*
Obbia *Spizocorys obbiensis*
masked *S. personata*
Dunn's *Eremalauda dunni*
bar-tailed *Ammomanes cincturus*
desert *A. deserti*
Gray's *A. grayi*
short-tailed *Pseudalaemon fremantlii*
lesser hoopoe *Alaemon hamertoni*
greater hoopoe *A. alaudipes*
Dupont's *Chersophilus duponti*
thick-billed *Ramphocoris clotbey*
Woodlark *Lullula arborea*
Skylark, Eurasian *Alauda arvensis*
Lark, raso *A. razae* **E**
shore horned *Eremophila alpestris*
Temminck's horned *E. bilopha*
Finchlark, chestnut-backed *Eremopterix leucotis*
grey-backed *E. verticalis*
black-eared *E. australis*
Fischer's *E. leucopareia*
chestnut-headed *E. signata*
black-crowned *E. nigriceps*

FAMILY: HIRUNDINIDAE (SWALLOWS, MARTINS)

Swallow, barn *Hirundo rustica*
grey-rumped *H. griseopyga*
mosque *H. senegalensis*
west African *H. domicella*
greater striped *H. cucullata*
red-breasted *H. semirufa*
lesser striped *H. abyssinica*
wire-tailed *H. smithii*
Angola *H. angolensis*
blue *H. atrocaerulea*
pearl-breasted *H. dimidiata*
white-throated *H. albigularis*
red-rumped *H. daurica*
South African *H. spilodera*
Ethiopian *H. aethiopica*
red-chested *H. lucida*
white-tailed *H. megaensis* **R**
white-throated blue *H. nigrita*
Red Sea *H. perdita*
forest *H. fuliginosa*
Preuss's *H. preussi*
red-throated *H. rufigula*
black & rufous *H. nigrorufa*
pied-winged *H. leucosoma*
Martin, rock *H. fuligula*
Eurasian crag *H. rupestris*
pale crag *H. obsoleta*
northern house *Delichon urbica*
banded *Riparia cincta*
plain *R. paludicola*
sand *R. riparia*
Congo *R. congica*
mascarene *Phedina borbonica*
Brazza's *P. brazzae*
African river *Pseudochelidon eurystomina*
Saw-wing, black *Psalidoprocne holomelas*
blue *P. pristoptera*

brown *P. antinorii*
Ethiopian *P. oleaginea*
eastern *P. orientalis*
white-headed *P. albiceps*
square-tailed *P. nitens*
Shari *P. chalybea*
Petit's *P. petiti*
Mangbettu *P. mangbettorum*
Fanti *P. obscura*
mountain *P. fuliginosa*

FAMILY: CAMPEPHAGIDAE (CUCKOO-SHRIKES)
TRIBE: ORIOLINI

Cuckoo-shrike, grey *Coracina caesia*
white-breasted *C. pectoralis*
Grauer's *C. graueri*
ashy *C. cinerea*
blue *C. azurea*
black *Campephaga flava*
red-shouldered *C. phoenicea*
Petit's *C. petiti*
purple-throated *C. quiscalina*
Ghana *C. lobata* **V**
oriole *Lobotos oriolina*

FAMILY: DICRURIDAE (DRONGOS)
TRIBE: DICRURINI

Drongo, fork-tailed *Dicrurus adsimilis*
square-tailed *D. ludwigii*
Comoro *D. fuscipennis* **R**
Mayotte *D. waldeni* **R**
velvet-mantled *D. modestus*
shining *D. atripennis*
crested *D. forficatus*

FAMILY: ORIOLIDAE (ORIOLES)
TRIBE: ORIOLINI

Oriole, Eurasian golden *Oriolus oriolus*
African golden *O. auratus*
São Tomé *O. crassirostris*
green-headed *O. chlorocephalus*
black-tailed *O. percivali*
African black-headed *O. larvatus*
western black-headed *O. brachyrhynchus*
dark-headed *O. monacha*
black-winged *O. nigripennis*

FAMILY: CORVIDAE (CROWS, RAVENS)
TRIBE: CORVINI

Crow, pied *Corvus albus*
house *C. splendens*
Cape *C. capensis*
carrion (hooded) *C. corone*
Raven *C. corax*
white-necked *C. albicollis*
brown-necked *C. ruficollis*
thick-billed *C. crassirostris*
fan-tailed *C. rhipidurus*
Jackdaw, Eurasian *C. monedula*
Rook *C. frugilegus*
Crow, Stresemann's bush *Zavattariornis stresemanni* **R**
Piapiac *Ptilostomus afer*
Chough *Pyrrhocorax pyrrhocorax*
Alpine *P. graculus*

Jay, Eurasian *Garrulus glandarius*
Magpie *Pica pica*

FAMILY: PARIDAE (TITS)
Subfamily: Parinae

Tit, grey *Parus afer*
white-shouldered *P. guineensis*
ashy *P. cinerascens*
miombo *P. griseiventris*
Carp's *P. carpi*
black *P. niger*
rufous-bellied *P. rufiventris*
white-bellied *P. albiventris*
Somali *P. thruppi*
cinnamon-breasted *P. pallidiventris*
stripe-breasted *P. fasciiventer*
red-throated *P. fringillinus*
white-winged *P. leucomelas*
white-backed *P. leuconotos*
dusky *P. funereus*
crested *P. cristatus*
coal *P. ater*
blue *P. caeruleus*
great *P. major*
long-tailed *Aegithalos caudatus*
hylia *Pholidornis rushiae*

FAMILY: REMICIDAE (PENDULINE TITS)
Subfamily: Remicinae

Penduline tit, African *Anthoscopus caroli*
mouse-coloured *A. musculus*
Sennar *A. punctifrons*
Cape *A. minutus*
yellow *A. parvulus*
forest *A. flavifrons*
buff-bellied *A. sylviella*
Eurasian *Remiz pendulinus*

FAMILY: SITTIDAE (NUTHATCHES)

Nuthatch, wood *Sitta europaea*
Kabylian *S. ledanti* **R**

FAMILY: CERTHIIDAE (TREE CREEPERS)

Tree creeper, short-toed *Certhia brachydactyla*
Eurasian *C. familiaris*

FAMILY: TICHODROMADIDAE (WALLCREEPER)

Wallcreeper *Tichodroma muraria*

FAMILY: SALPORNITHIDAE (SPOTTED CREEPER)

Creeper, spotted *Salpornis spilonotus*

FAMILY: TIMALIIDAE (BABBLERS)
Subfamily: Sylviinae

Spot-throat *Modulatrix stictigula*
Dapple-throat *Arcanator orostruthus*
Babbler, arrow-marked *Turdoides jardineii*
Angola *T. hartlaubii*
black-lored *T. melanops*
southern pied *T. bicolor*
northern pied *T. hypoleucus*
bare-cheeked *T. gymnogenys*
Cretschmar's *T. leucocephalus*

common *T. caudatus*
dusky *T. tenebrosus*
brown *T. plebejus*
white-rumped *T. leucopygius*
Hinde's pied *T. hindei* **V**
scaly *T. squamulatus*
blackcap *T. reinwardti*
Chatterer, rufous *T. rubiginosus*
scaly *T. aylmeri*
fulvous *T. fulvus*
Babbler, thrush *Ptyrticus turdinus*
capuchin *Phyllanthus atripennis*
white-throated mountain *Lioptilus gilberti* **R**
blackcap mountain *L. nigricapillus*
Chapin's mountain *Kupeornis chapini*
red-collared mountain *K. rufocinctus*
Abyssinian hill *Illadopsis abyssinica*
Ruwenzori hill *I. atriceps*
Illadopsis, blackcap *I. cleaveri*
scaly-breasted *I. albipectus*
pale-breasted *I. rufipennis*
mountain *I. pyrrhoptera*
brown *I. fulvescens*
Puvel's *I. puveli*
rufous-winged *I. rufescens*
grey-chested *Kakamega poliothorax*
Catbird, Abyssinian *Parophasma galinieri*
Leiothrix, red-billed (Pekin robin) *Leiothrix lutea*

FAMILY: PYCNONOTIDAE (BULBULS, GREENBULS)

Bulbul, common (garden) *Pycnonotus barbatus*
white-spectacled *P. xanthopygos*
Cape *P. capensis*
black-fronted *P. nigricans*
Dodson's *P. dodsoni*
dark-capped *P. tricolor*
Somali *P. somaliensis*
white-bearded *Criniger ndussumensis*
green-backed *C. chloronotus*
red-tailed *C. calurus*
yellow-bearded *C. olivaceus* **V**
bearded *C. barbatus*
black-collared *Neolestes torquatus*
Comoro *Hypsipetes parvirostris*
Greenbul, little *Andropadus virens*
grey *A. gracilis*
Ansorge's *A. ansorgei*
plain *A. curvirostris*
slender-billed *A. gracilirostris*
grey-throated *A. tephrolaemus*
stripe-cheeked *A. milanjensis*
Shelley's *A. masukuensis*
yellow-whiskered *A. latirostris*
sombre *A. importunus*
Cameroon *A. montanus*
Kakamega *A. kakamegae*
Hall's *A. hallae*
mountain *A. nigriceps*
green-throated *A. chlorigula*
olive-headed *A. olivaceiceps*
swamp *Thescelocichla leucopleura*
yellow-throated *Chlorocichla flavicollis*
simple *C. simplex*

yellow-bellied *C. flaviventris*
joyful *C. laetissima*
Prigogine's *C. prigoginei* **V**
yellow-necked *C. falkensteini*
white-tailed *Baeopogon clamans*
honeyguide *B. indicator*
spotted *Ixonotus guttatus*
golden *Calyptocichla serina*
Bristlebill, common *Bleda syndactyla*
  green-tailed *B. eximia*
  grey-headed *B. canicapilla*
Greenbul, Fischer's *Phyllastrephus fischeri*
  Cabanis's *P. cabanisi*
  grey-olive *P. cerviniventris*
  Sassi's *P. lorenzi*
  placid *P. placidus*
  yellow-streaked *P. flavostriatus*
  Sharpe's *P. alfredi*
  Xavier's *P. xavieri*
  icterine *P. icterinus*
  white-throated *P. albigularis*
  tiny *P. debilis*
  pale-olive *P. fulviventris*
  Liberian *P. leucolepis*
  grey-headed *P. poliocephalus*
Brownbul, terrestrial *P. terrestris*
  northern *P. strepitans*
Olive-greenbul, Toro *P. hypochloris*
  Baumann's *P. baumanni*
  Cameroon *P. poensis*
Leaf-love *P. scandens*
Nicator, eastern *Nicator gularis*
  yellow-spotted *N. chloris*
  yellow-throated *N. vireo*

FAMILY: BOMBYCILLIDAE (WAXWINGS, GREY HYPOCOLIUS)

Waxwing, Bohemian *Bombycilla garrulus*
Hypocolius, grey *Hypocolius ampelinus*

FAMILY: CINCLIDAE (DIPPERS)

Dipper, white-throated *Cinclus cinclus*

FAMILY: TROGLODYTIDAE (WRENS)

Wren *Troglodytes troglodytes*

FAMILY: PRUNELLIDAE (ACCENTORS)

Accentor, hedge *Prunella modularis*
  alpine *P. collaris*

FAMILY: TURDIDAE/MUSCICAPIDAE (THRUSHES, CHATS, ROBINS)

Subfamily: Turdinae

Thrush, Comoro *Turdus bewsheri*
  taita *T. helleri* **E**
  orange *T. (Zoothera) gurneyi*
  olive *T. olivaceus*
  groundscraper *T. (Psophocichla) litsitsirupa*
  Kurrichane *T. libonyanus*
  spotted *T. fischeri* **R**
  olivaceus *T. olivaceofuscus*
  African *T. pelios*
  song *T. philomelos*
  bare-eyed *T. tephronotus*

black-throated *T. ruficollis*
mistle *T. viscivorus*
Ouzel, ring *T. torquatus*
Blackbird, Eurasian *T. merula*
Fieldfare *T. pilaris*
Redwing *T. iliacus*
Flycatcher-thrush, Finsch's *Neocossyphus finschi*
  rufous *N. fraseri*
Ant thrush, white-tailed *N. poensis*
  red-tailed *N. rufus*
Rock thrush, Cape *Monticola rupestris*
  sentinel *M. explorator*
  Transvaal *M. pretoriae*
  short-toed *M. brevipes*
  miombo *M. angolensis*
  little *M. rufocinereus*
  rufous-tailed *M. saxatilis*
  blue *M. solitaria*

TRIBE: SAXICOLINI

Chat, buff-streaked *Saxicola bifasciata*
Whinchat *S. rubetra*
Stonechat *S. torquata*
Chat, Canary Islands *S. dacotiae* **R**
Wheatear, capped *Oenanthe pileata*
  northern *O. oenanthe*
  pied *O. pleschanka*
  mountain *O. monticola*
  isabelline *O. isabellina*
  black-eared *O. hispanica*
  desert *O. deserti*
  hooded *O. monacha*
  Botta's *O. bottae*
  mourning *O. lugens*
  rufous-tailed *O. xanthoprymna*
  white-tailed *O. leucopyga*
  Cyprus *O. cypriaca*
  Hume's *O. alboniger*
  Schalow's *O. lugubris*
  Heuglin's *O. heuglini*
  Somali *O. phillipsi*
  Finsch's *O. finschii*
  red-rumped *O. moesta*
  black *O. leucura*
Chat, familiar *Cercomela familiaris*
  sickle-winged *C. sinuata*
  Karoo *C. schlegelii*
  tractrac *C. tractrac*
  brown-tailed *C. scotocerca*
  sombre *C. dubia*
  moorland *C. sordida*
Blackstart *C. melanura*
Cliff chat, white-crowned *Thamnolaea coronata*
  white-winged *T. semirufa*
  mocking *T. cinnamomeiventris*
Black-chat, white-headed *Myrmecocichla arnotti*
Anteater-chat, southern *M. formicivora*
Black-chat, white-fronted *M. albifrons*
Chat, Rüppell's *M. melaena*
Anteater-chat, northern *M. aethiops*
Chat, sooty *M. nigra*
Moor-chat, Congo *M. tholloni*
Scrub-robin, forest *Cercotrichas leucosticta*

brown-backed *C. hartlaubi*
rufous-tailed *C. galactotes*
miombo *C. barbata*
black *C. podobe*
brown *Erythropygia signata*
bearded *E. quadrivirgata*
Karoo *E. coryphaeus*
Kalahari *E. paena*
red-backed *E. leucophrys*
Robin-chat, chorister *Cossypha dichroa*
  red-capped *C. natalensis*
  white-browed *C. heuglini*
  Cape *C. caffra*
  white-throated *C. humeralis*
  grey-winged *C. polioptera*
  Archer's *C. archeri*
  olive-flanked *C. anomala*
  blue-shouldered *C. cyanocampter*
  Rüppell's *C. semirufa*
  snowy-crowned *C. niveicapilla*
  white-crowned *C. albicapilla*
  white-headed *C. heinrichi*
  mountain *C. isabellae*
  white-bellied *C. roberti*
Robin, white-starred *Pogonocichla stellata*
  Swynnerton's *Swynnertonia swynnertoni* **R**
  forest *Stiphrornis erythrothorax*
Akalat, east coast *Sheppardia gunningi* **R**
  Usambara *S. montana*
  Iringa *S. lowei* **R**
  Sharpe's *S. sharpei*
  lowland *S. cyornithopsis*
  equatorial *S. aequatorialis*
  Gabela *S. gabela*
  Bocage's *S. bocagei*
  Alexander's *S. poensis*
Nightingale, thrush *Luscinia luscinia*
  common *L. megarhynchos*
Bluethroat *L. svecica*
Alethe, white-chested *Alethe fuelleborni*
  fire-crested *A. castanea*
  red-throated *A. poliophrys*
  white-tailed *A. diademata*
  brown-chested *A. poliocephala*
  Cholo *A. choloensis* **R**
Palm thrush, collared *Cichladusa arquata*
  rufous-tailed *C. ruficauda*
Thrush, spotted morning *C. guttata*
Redstart, common *Phoenicurus phoenicurus*
  black *P. ochruros*
  Moussier's *P. moussieri*
Robin, white-throated *Irania gutturalis*
  European *Erithacus rubecula*
Ground robin, Usambara *Dryocichloides montanus* **R**
Cave chat, Angola *Xenocopsychus ansorgei*
Chat, boulder *Pinarornis plumosus*
  Herero *Namibornis herero*
Ground thrush, spotted *Zoothera guttata*
  black-eared *Z. cameronensis*
  Kibale *Z. kibalensis*
  Abyssinian *Z. piaggiae*
  Oberlaender's *Z. oberlaenderi* **R**
  Kivu *Z. tanganjicae*
  Crossley's *Z. crossleyi*

grey *Z. princei*

FAMILY: PICATHARTIDAE

Rockfowl, grey-necked *Picathartes oreas* **R**
 white-necked *P. gymnocephalus* **V**
Rockjumper, rufous *Chaetops frenatus*
 orange-breasted *C. aurantius*

FAMILY: SYLVIIDAE (WARBLERS)

Warbler, garden *Sylvia borin*
 Red Sea *S. leucomelaena*
 subalpine *S. cantillans*
 Rüppell's *S. rueppelli*
 Ménétries's *S. mystacea*
 Orphean *S. hortensis*
 barred *S. nisoria*
 desert *S. nana*
 Marmora's *S. sarda*
 Dartford *S. undata*
 Tristram's *S. deserticola*
 spectacled *S. conspicillata*
 Sardinian *S. melanocephala*
 Cyprus *S. melanothorax*
Whitethroat *S. communis*
 lesser *S. curruca*
Blackcap *S. atricapilla*
Crombec, red-capped *Sylvietta ruficapilla*
 Cape *S. rufescens*
 red-faced *S. whytii*
 white-browed *S. leucophrys*
 Chapin's *S. chapini*
 green *S. virens*
 northern *S. brachyura*
 short-billed *S. philippae*
 Somali *S. isabellina*
 lemon-bellied *S. denti*
Warbler, olive-tree *Hippolais olivetorum*
 icterine *H. icterina*
 Upcher's *H. languida*
 olivaceous *H. pallida*
 melodious *H. polyglotta*
 willow *Phylloscopus trochilus*
 wood *P. sibilatrix*
 Bonelli's *P. bonelli*
 yellow-browed *P. inornatus*
 dusky *P. fuscatus*
Woodland warbler, yellow-throated *P. ruficapilla*
 Laura's *P. laurae*
 red-faced *P. laetus*
 Uganda *P. budongensis*
 brown *P. umbrovirens*
 black-capped *P. herberti*
Chiffchaff *P. collybita*
Warbler, yellow-throated *Seicercus ruficapillus*
Scrub warbler, Knysna *Bradypterus sylvaticus*
 Victorin's *B. victorini*
 African *B. barratti*
 Cameroon *B. lopezi*
 white-winged *B. carpalis*
 Dja River *B. grandis*
Warbler, African bush *B. baboecala*
 Grauer's rush *B. graueri* **V**
 cinnamon bracken *B. cinnamomeus*
 evergreen forest *B. mariae*

bamboo *B. alfredi*
Warbler, Eurasian river *Locustella fuviatilis*
 common grasshopper *L. naevia*
 Savi's *L. luscinoides*
Grassbird, fan-tailed *Schoenicola brevirostris*
Flycatcher-warbler, yellow *Chloropeta natalensis*
 mountain *C. similis*
 thin-billed *C. gracilirostris* **R**
Warbler, fairy *Stenostira scita*
Reed warbler, great *Acrocephalus arundinaceus*
 African *A. baeticatus*
 Eurasian *A. scirpaceus*
Warbler, marsh *A. palustris*
 greater swamp *A. rufescens*
 sedge *A. schoenobaenus*
 lesser swamp *A. gracilirostris*
 moustached *A. melanopogon*
 clamorous reed *A. stentoreus*
 aquatic *A. paludicola*
Swamp warbler, Cape Verde *A. brevipennis*
Warbler, moustached grass *Melocichla mentalis*
Hyliota, yellow-bellied *Hyliota flavigaster*
 southern *H. australis*
 violet-backed *H. violacea*
Warbler, Socotra *Incana incana*
 buff-bellied *Phyllolais pulchella*
 Neumann's *Hemitesia neumanni*
 Grauer's *Graueria vittata*
 Cetti's *Cettia cetti*
Brush warbler, Moheli *Nesillas mariae*
 Anjouan *N. longicaudata*
 Grand Comoro *N. brevicaudata*
Warbler, white-tailed *Poliolais lopezi*
 Mrs Moreau's *Scypomycter winifreda* **R**
 black-capped rufous *Bathmocercus cerviniventris*
 black-faced rufous *B. rufus*
Tailorbird, long-billed *Orthotomus moreaui*
 African *O. metopias*
Warbler, scrub *Scotocerca inquieta*
 grey-capped *Eminia lepida*
 oriole *Hypergerus atriceps*
 cricket *Spiloptila clamans*
 red-fronted *S. rufifrons*
 grey bush *Calamonastes simplex*
 miombo bush *C. undosus*
Wren-warbler, barred *C. fasciolatus*
 Stierling's *C. stierlingi*
Warbler, cinnamon-breasted *Euryptila subcinnamomea*
 rufous-eared *Malcorus pectoralis*
 red-winged *Heliolais erythroptera*
 red-winged grey *Drymocichla incana*
 brier *Oreophilais robertsi*
Camaroptera, green-backed *Camaroptera brachyura*
 yellow-browed *C. superciliaris*
 olive-green *C. chloronota*
 grey-backed *C. brevicaudatus*
 Hartert's *C. harterti*
Prinia, Namaqua *Phragmacia substriata*
 graceful *Prinia gracilis*
 spotted *P. maculosa*
 tawny-flanked *P. subflava*

black-chested *P. flavicans*
river *P. fluviatilis*
white-chinned *P. leucopogon*
pale *P. somalica*
banded *P. bairdii*
black-faced *P. melanops*
Sierra Leone *P. leontica*
São Tomé *P. molleri*
Cisticola, Lepe *Cisticola lepe*
 Lynes's *C. distinctus*
 desert *C. aridulus*
 wing-snapping *C. ayresii*
 brown-backed *C. discolor*
 rock-loving *C. emini*
 Mongalla *C. mongalla*
 cloud *C. textrix*
 Dorst's *C. dorsti*
 grey *C. rufilatus*
 lazy *C. aberrans*
 rattling *C. chinianus*
 red-headed *C. subruficapillus*
 croaking *C. natalensis*
 Angola *C. angolensis*
 wailing *C. lais*
 Tabora *C. angusticauda*
 red-faced *C. erythrops*
 singing *C. cantans*
 chirping *C. pipiens*
 tinkling *C. tinniens*
 pectoral-patch *C. brunnescens*
 black-necked *C. eximius*
 zitting *C. juncidis*
 stout *C. robustus*
 Aberdare *C. aberdare*
 ashy *C. cinereolus*
 Tana River *C. restrictus*
 churring *C. njombe*
 red-pate *C. ruficeps*
 tiny *C. nanus*
 siffling *C. brachypterus*
 foxy *C. troglodytes*
 whistling *C. lateralis*
 trilling *C. woosnami*
 Hunter's *C. hunteri*
 Chubb's *C. chubbi*
 black-tailed *C. melanurus*
 black-lored *C. nigriloris*
 Boran *C. bodessa*
 winding *C. galactotes*
 Carruthers's *C. carruthersi*
 cloud-scraping *C. dambo*
 rufous *C. rufus*
 chattering *C. anonymus*
 bubbling *C. bulliens*
 island *C. haesitatus*
 Socotra *C. incanus*
Neddicky (Piping cisticola) *C. fulvicapillus*
Apalis, black-headed *Apalis melanocephala*
 Chirinda *A. chirindensis*
 yellow-breasted *A. flavida*
 bar-throated *A. thoracica*
 black-faced *A. personata*
 brown-tailed *A. viridiceps*
 Gosling's *A. goslingi*
 Rudd's *A. ruddi*
 masked *A. binotata*

black-throated *A. jacksoni*
white-winged *A. chariessa*
black-capped *A. nigriceps*
black-collared *A. pulchra*
collared *A. ruwenzorii*
grey *A. cinerea*
brown-headed *A. alticola*
Bamenda *A. bamendae*
red-faced *A. rufifrons*
Karamoja *A. karamojae*
buff-throated *A. rufogularis*
Kungwe *A. argentea* **R**
chestnut-throated *A. porphyrolaema*
Chapin's *A. chapini*
Kabobo *A. kaboboensis* **R**
Warbler, Pearson's *A. melanura*
Apalis, Sharpe's *A. sharpii*
long-billed *A. moreaui* **R**
Eremomela, burnt-neck *Eremomela usticollis*
green-capped *E. scotops*
yellow-bellied *E. icteropygialis*
yellow-rumped *E. gregalis*
yellow-vented *E. flavicrissalis*
rufous-crowned *E. badiceps*
green-backed *E. canescens*
Turner's *E. turneri* **R**
Salvadori's *E. salvadorii*
Senegal *E. pusilla*
black-necked *E. atricollis*
Grass warbler, Cape *Sphenoeacus afer*
Rockrunner, Dama *Achaetops pycnopygius*
Titbabbler *Parisoma subcaeruleum*
Layard's *P. layardi*
Parisoma, brown *P. lugens*
banded *P. boehmi*
Hylia, green *Hylia prasina*
Longbill, grey *Macrosphenus concolor*
Kretschmer's *M. kretschmeri*
yellow *M. flavicans*
Pulitzer's *M. pulitzeri*
Kemp *M. kempi*
Bocage's *Amaurocichla bocagii*
Longtail, green *Urolais epichlora*
Goldcrest *Regulus regulus*
Firecrest *R. ignicapillus*
Kinglet, Tenerife *R. teneriffae*

FAMILY: MUSCICAPIDAE (FLYCATCHERS, BATISES)

Flycatcher, spotted *Muscicapa striata*
sooty *M. infuscata*
Gambaga *M. gambagae*
dusky blue *M. comitata*
Böhm's *M. boehmi*
Ussher's *M. ussheri*
Tessmann's *M. tessmanni*
Alseonax, blue-grey (ashy) *M. caerulescens*
dusky *M. adusta*
Itombwe *M. itombwensis*
Chapin's *M. lendu* **R**
swamp *M. aquatica*
Cassin's *M. cassini*
yellow-footed *M. sethsmithii*
olivaceus *M. olivascens*
little grey *M. epulata*
Flycatcher, collared *Ficedula albicollis*
European pied *F. hypoleuca*

red-breasted *F. parva*
semi-collared *F. semitorquata*
grey-throated *Myioparus griseigularis*
Tit-flycatcher, great *M. plumbeus*
Flycatcher, pale *Bradornis pallidus*
Marico *B. mariquensis*
chat *B. infuscatus*
little grey *B. pumilus*
large *B. microrhychus*
Silverbird *B. semipartitus*
Black flycatcher, southern *Melaenornis pammelaina*
northern *M. edolioides*
yellow-eyed *M. ardesiacus*
west African *M. annamarulae*
Slaty flycatcher, white-eyed *Dioptornis fischeri*
Abyssinian *D. chocolatinus*
Angola *D. brunneus*
Flycatcher, fiscal *Sigelus silens*
Livingstone's *Erythrocercus livingstonei*
chestnut-capped *E. mccallii*
Monarch, yellow *E. holochlorus*
Flycatcher, blue-mantled crested *Trochocercus cyanomelas*
white-tailed crested *T. albonotatus*
white-bellied crested *T. albiventris*
blue-headed crested *T. nitens*
dusky *T. nigromitratus*
Dorhn's *Horizorhinus dohrni*
Grand Comoro *Humblotia flavirostris* **R**
forest *Fraseria ocreata*
white-browed forest *F. cinerascens*
Paradise flycatcher, rufous-vented *Terpsiphone rufocinerea*
Madagascar *T. mutata*
African *T. viridis*
black-headed *T. rufiventer*
São Tomé *T. atrochalybea*
Bedford's *T. bedfordi*
Flycatcher, African blue *Elminia longicauda*
white-tailed blue *E. albicauda*
Shrike-flycatcher, black & white *Bias musicus*
African *B. flammulatus*
Wattle-eye, black-throated *Platysteira peltata*
brown-throated *P. cyanea*
chestnut *P. castanea*
Jameson's *P. jamesoni*
yellow-bellied *P. concreta*
white-fronted *P. albifrons*
banded *P. laticincta* **E**
white-spotted *P. tonsa*
red-cheeked *P. blissetti*
black-necked *P. chalybea*
Batis, Cape *Batis capensis*
pririt *B. pririt*
chinspot *B. molitor*
Zululand *B. fratrum*
pale *B. soror*
short-tailed *B. mixta*
Ruwenzori *B. diops*
grey-headed *B. orientalis*
pygmy *B. perkeo*
black-headed *B. minor*
Ituri *B. ituriensis*
Boulton's *B. margaritae*
Verreaux's *B. minima*

Senegal *B. senegalensis*
Fernando Po *B. poensis*
Angola *B. minulla*
Malawi *B. dimorpha*
Reichenow's *B. reichenowi*
west African *B. occultus*

FAMILY: MOTACILLIDAE (WAGTAILS, PIPITS, LONGCLAWS)

Wagtail, African pied *Motacilla aguimp*
Cape *M. capensis*
mountain *M. clara*
yellow *M. flava*
grey *M. cinerea*
citrine *M. citreola*
white *M. alba*
Pipit, mountain *Anthus hoeschi*
buffy *A. vaalensis*
plain-backed *A. leucophrys*
African *A. cinnamomeus*
long-billed *A. similis*
rock (yellow-tufted) *A. crenatus*
short-tailed *A. brachyurus*
tree *A. trivialis*
bush *A. caffer*
red-throated *A. cervinus*
striped *A. lineiventris*
Cameroon *A. cameroonensis*
Bannerman's *A. bannermani*
Jackson *A. latistriatus*
yellow-breasted *A. chloris*
Richard's *A. novaeseelandiae*
tawny *A. campestris*
meadow *A. pratensis*
water *A. spinoletta*
Sokoke *A. sokokensis*
Malindi *A. melindae*
long-legged *A. pallidiventris*
rock *A. petrosus*
woodland *A. nyassae*
Sharpe's *A. sharpei*
long-tailed *A. longicandatus*
golden *Tmetothylacus tenellus*
Longclaw, rosy-throated *Macronyx amelia*
yellow-throated *M. croceus*
Cape *M. capensis*
Fülleborn's *M. fuellebornii*
Pangani *M. aurantiigula*
Abyssinian *M. flavicollis*
Grimwood's *M. grimwoodi*

FAMILY: LANIIDAE (SHRIKES)

Shrike, red-backed *Lanius collurio*
Souza's *L. souzae*
lesser grey *L. minor*
rufous-tailed *L. isabellinus*
masked *L. nubicus*
woodchat *L. senator*
Mackinnon's *L. mackinnoni*
great grey *L. excubitor*
Emin's *L. gubernator*
Fiscal, common *L. collaris*
Somali *L. somalicus*
taita *L. dorsalis*
grey-backed *L. excubitoroides*
long-tailed *L. cabanisi*

Uhehe *L. marwitzi*
Newton's *L. newtoni*
Shrike, brown *L. cristatus*
　magpie *Corvinella melanoleuca*
　yellow-billed *C. corvina*
　white-crowned *Eurocephalus anguitimens*
　white-rumped *E. rueppelli*

FAMILY: MALACONOTIDAE (BUSH SHRIKES, TCHAGRAS)

Bush shrike, grey-headed *Malaconotus blanchoti*
　fiery-breasted *M. cruentus*
　Uluguru *M. alius* **R**
　Lagden's *M. lagdeni*
　Perrin's *M. viridis*
　green-breasted *M. gladiator* **R**
　Monteiro's *M. monteiri*
　grey-green *Telophorus bocagei*
　many-coloured *T. multicolor*
　sulphur-breasted *T. sulfureopectus*
　Doherty's *T. dohertyi*
　Mount Kupe *T. kupeensis*
　four-coloured *T. quadricolor*
　black-fronted *T. nigrifrons*
　olive *T. olivaceus*
　sulphur-breasted *T. sulfureopectus*
Bokmakierie *T. zeylonus*
Gonolek, crimson-breasted *Laniarius atrococcineus*
　black-headed *L. erythrogaster*
　papyrus *L. mufumbiri*
　common *L. barbarus*
Boubou, tropical *L. aethiopicus*
　southern *L. ferrugineus*
　Gabela *L. amboinensis*
　Gabon *L. bicolor*
　slate-coloured *L. funebris*
　Fülleborn's *L. fuelleborni*
　sooty *L. leucorhynchus*
　Bulo Burti *L. liberatus*
　mountain *L. poensis*
　yellow-breasted *L. atroflavus*
　Turati's *L. turatii*
Bush shrike, orange-breasted *L. brauni*
　red-naped *L. ruficeps*
　Luehder's *L. luehderi*
Shrike, chat *Lanioturdus torquatus*
Brubru *Nilaus afer*
Bush shrike, rosy-patched *Rhodophoneus cruentus*
Puffback, black-backed *Dryoscopus cubla*
　northern *D. gambensis*
　Pringle's *D. pringlii*
　red-eyed *D. senegalensis*
　pink-footed *D. angolensis*
　large-billed *D. sabini*
Tchagra, brown-crowned *Tchagra australis*
　Anchieta's *T. anchietae*
　black-crowned *T. senegala*
　southern *T. tchagra*
　marsh *T. minuta*
　three-streaked *T. jamesi*

FAMILY: PRIONOPIDAE (HELMETSHRIKES)

Helmetshrike, white *Prionops plumatus*
　Gabon *P. rufiventris*
　chestnut-fronted *P. scopifrons*
　Retz's *P. retzii*
　yellow-crested *P. alberti*
　grey-crested *P. poliolophus*
　Angola *P. gabela*
　chestnut-bellied *P. caniceps*

FAMILY: STURNIDAE (STARLINGS, MYNAS)

Starling, common *Sturnus vulgaris*
　spotless *S. unicolor*
　rosy *S. roseus*
　African pied *Spreo bicolor*
　Fischer's *S. fischeri*
　white-crowned *S. albicapillis*
Grackle, Tristram's *Onychognathus tristramii*
Starling, pale-winged *O. nabouroup*
　red-winged *O. morio*
　bristle-crowned *O. salvadorii*
　slender-billed *O. tenuirostris*
　chestnut-winged *O. fulgidus*
　Somali *O. blythii*
　Socotra *O. frater*
　Waller's *O. walleri*
　white-billed *O. albirostris*
　wattled *Creatophora cinerea*
　chestnut-bellied *Lamprotornis pulcher*
　Shelley's *L. shelleyi*
　Hildebrandt's *L. hildebrandti*
　superb *L. superbus*
Glossy starling, red-shouldered *L. nitens*
　greater blue-eared *L. chalybaeus*
　lesser blue-eared *L. chloropterus*
　southern blue-eared *L. elisabeth*
　black-bellied *L. corruscus*
　sharp-tailed *L. acuticaudus*
　Meves's *L. mevesii*
　Burchell's *L. australis*
　bronze-tailed *L. chalcurus*
　Rüppell's *L. purpuropterus*
　splendid *L. splendidus*
　purple *L. purpureus*
　purple-headed *L. purpureiceps*
　copper-tailed *L. cupreocauda*
　long-tailed *L. caudatus*
　Principe *L. ornatus*
　emerald *Coccycolius iris*
Starling, plum-coloured *Cinnyricinclus leucogaster*
　Sharpe's *C. sharpii*
　Abbot's *C. femoralis*
　narrow-tailed *Poeoptera lugubris*
　Stuhlmann's *P. stuhlmanni*
　Kenrick's *P. kenricki*
　white-collared *Grafisia torquata*
　magpie *Speculipastor bicolor*
　babbling *Neocichla gutturalis*
　golden-breasted *Cosmopsarus regius*
　ashy *C. unicolor*
Myna, common *Acridotheres tristis*

FAMILY: BUPHAGIDAE (OXPECKERS)

Oxpecker, yellow-billed *Buphagus africanus*
　red-billed *B. erythrorhynchus*

FAMILY: NECTARINIIDAE (SUNBIRDS)

Subfamily: Promeropinae (sugarbirds)

Sugarbird, Cape *Promerops caffer*
　Gurney's *P. gurneyi*

TRIBE: NECTARINIINI

Sunbird, malachite *Nectarinia famosa*
　orange-breasted *N. violacea*
　bronze *N. kilimensis*
　Newton's *N. newtonii*
　Shelley's *N. shelleyi*
　scarlet-chested *N. senegalensis*
　olive *N. olivacea*
　Johanna's *N. johannae*
　mouse-coloured *N. veroxii*
　dusky *N. fusca*
　white-bellied *N. talatala*
　variable *N. venusta*
　long-billed green *N. notata*
　Neergaard's *N. neergaardi*
　Marico (Mariqua) *N. mariquensis*
　greater double-collared *N. afra*
　purple-banded *N. bifasciata*
　southern double-collared *N. chalybea*
　miombo double-collared *N. manoensis*
　Ursula's *N. ursulae*
　Oustalet's *N. oustaleti*
　orange-tufted *N. bouvieri*
　shining *N. habessinica*
　violet-breasted *N. pembae*
　splendid *N. coccinigastra*
　superb *N. superba*
　olive-bellied *N. chloropygia*
　tiny *N. minulla*
　montane double-collared *N. ludovicensis*
　northern double-collared *N. preussi*
　regal *N. regia*
　eastern double-collared *N. mediocris*
　Loveridge's *N. loveridgei*
　Moreau's *N. moreaui*
　São Tomé *N. thomensis*
　Stuhlmann's double-collared *N. stuhlmanni*
　red-chested *N. erythrocerca*
　black-bellied *N. nectarinioides*
　beautiful *N. pulchella*
　little green *N. seimundi*
　rufous-winged *N. rufipennis* **R**
　copper *N. cuprea*
　purple-breasted *N. purpureiventris*
　Tacazze *N. tacazze*
　scarlet-tufted *N. johnstoni*
　golden-winged *N. reichenowi*
　Humblot's *N. humbloti*
　Anjouan *N. comorensis*
　Mayotte *N. coquerellii*
　blue-throated brown *N. cyanolaema*
　green-headed *N. verticalis*
　Bannerman's *N. bannermani*
　blue-headed *N. alinae*
　Cameroon *N. oritis*
　Hunter's *N. hunteri*
　amethyst *N. amethystina*
　green-throated *N. rubescens*
　Reichenbach's *N. reichenbachii*

Principe *N. harti*
Socotra *N. balfouri*
Prigogine's double-collared
    N. prigoginei **E**
Rockefeller's *N. rockefelleri* **R**
Palestine *N. osea*
Congo *N. congensis*
Bocage's *N. bocagei*
Bates's *N. batesi*
buff-throated *N. adelberti*
carmelite *N. fuliginosa*
mouse-brown *Anthreptes gabonicus*
plain-backed *A. reichenowi*
collared *A. collaris*
western violet-backed *A. longuemarei*
grey-headed *A. axillaris*
violet-tailed *A. aurantium*
Nile valley *A. metallicus*
pygmy *A. platurus*
green *A. rectirostris*
banded *A. rubritorques* **R**
Kenya violet-backed *A. orientalis*
Uluguru violet-backed *A. neglectus*
Anchieta's *A. anchietae*
scarlet-tufted *A. fraseri*
Amani *A. pallidigaster* **R**

FAMILY: ZOSTEROPIDAE (WHITE-EYES, SPEIROPS)

White-eye, African yellow *Zosterops senegalensis*
Cape (pale) *Z. pallidus*
white-breasted *Z. abyssinica*
broad-ringed *Z. poliogaster*
Principe *Z. ficedulinus*
Comoro *Z. mouroniensis* **R**
Pemba *Z. vaughani*
Kirk's *Z. madaraspatensis*
chestnut-sided *Z. mayottensis*
Speirops, Fernando Po *Speirops brunneus* **R**
black-capped *S. lugubris*
Principe *S. leucophaea*
Cameroon *S. melanocephalus*

FAMILY: PLOCEIDAE/PASSERIDAE (SPARROWS,
WEAVERS, BISHOPS, WIDOWS, QUELEAS)

Sparrow, house *Passer domesticus*
southern rufous *P. motitensis*
Spanish *P. hispaniolensis*
Iago *P. iagonensis*
Cape *P. melanurus*
Kenya rufous *P. rufocinctus*
grey-headed *P. griseus*
Socotra *P. insularis*
Somali *P. castanopterus*
Sudan golden *P. luteus*
Arabian golden *P. euchlorus*
chestnut *P. eminibey*
desert *P. simplex*
Swainson's *P. swainsonii*
parrot-billed *P. gongonensis*
Swahili *P. suahelicus*
Eurasian tree *P. montanus*
rock *Petronia petronia*
Petronia, yellow-throated *P. superciliaris*
yellow-spotted *P. pyrgita*
bush *P. dentata*
pale *Carpospiza brachydactyla*

Sparrow, Java *Padda oryzivora*
Weaver, village *Ploceus cucullatus*
Veillot's black *P. nigerrimus*
Speke's *P. spekei*
Fox's *P. spekeoides*
orange *P. aurantius*
northern brown-throated *P. castanops*
southern brown-throated *P. xanthopterus*
black-headed *P. melanocephalus*
cinnamon *P. badius*
Salvadori's *P. dichrocephalus*
golden-backed *P. jacksoni*
bar-winged *P. angolensis*
Rüppell's *P. galbula*
chestnut *P. rubiginosus*
little *P. luteolus*
slender-billed *P. pelzelni*
Kilombero *Ploceus burnieri*
Weyns's *P. weynsi*
Clarke's *P. golandi* **E**
forest *P. bicolor*
spectacled *P. ocularis*
black-necked *P. nigricollis*
black-billed *P. melanogaster*
strange *P. alienus*
brown-capped *P. insignis*
Baglafecht *P. baglafecht*
Usambara *P. nicolli* **R**
olive-headed *P. olivaceiceps*
yellow-mantled *P. tricolor*
Maxwell's black *P. albinucha*
Bertrand's *P. bertrandi*
spotted-backed *P. cucullatus*
chestnut *P. rubiginosus*
Cape *P. capensis*
streaked *P. manyar*
Bannerman's *P. bannermani* **V**
Bates's *P. batesi* **R**
black-chinned *P. nigrimentum*
Loango *P. subpersonatus*
golden-naped *P. aureonucha*
yellow-legged *P. flavipes* **V**
giant *P. grandis*
Bocage's *P. temporalis*
Preuss's *P. preussi*
yellow-capped *P. dorsomaculatus*
bar-winged *P. angolensis*
São Tomé *P. sanctithomae*
Masked weaver, northern *P. taeniopterus*
lesser *P. intermedius*
southern *P. velatus*
Tanzanian *P. reichardi*
Heuglin's *P. heuglini*
vitelline *P. vitellinus*
Katanga *P. katangae*
Victoria *P. victoriae*
Ruwet's *P. ruweti*
Golden weaver, Principe *P. princeps*
Taveta *P. castaneiceps*
palm *P. bojeri*
Holub's *P. xanthops*
eastern *P. subaureus*
African *P. subaureus*
Weaver, compact *Pachyphantes superciliosus*
Weaver, speckle-fronted *Sporopipes frontalis*
rufous-tailed *Histurgops ruficauda*

parasitic *Anomalospiza imberbis*
grosbeak *Amblyospiza albifrons*
sociable *Philetairus socius*
Sparrow-weaver, chestnut-backed *Plocepasser
    rufoscapulatus*
white-browed *P. mahali*
chestnut-crowned *P. superciliosus*
Donaldson-Smith's *P. donaldsoni*
Social weaver, grey-headed *Pseudonigrita
    arnaudi*
black-capped *P. cabanisi*
Buffalo weaver, white-billed *Bubalornis
    albirostris*
red-billed *B. niger*
white-headed *Dinemellia dinemelli*
Weaver, scaly *Sporopipes squamifrons*
red-headed *Anaplectes rubriceps*
Malimbe, red-bellied *Malimbus erythrogaster*
red-headed *M. rubricollis*
crested *M. malimbicus*
Gray's *M. nitens*
Ibadan *M. ibadanensis* **E**
Ballmann's *M. ballmanni*
red-vented *M. scutatus*
black-throated *M. cassini*
yellow-legged *M. flavipes*
Weaver, Rachel's *M. racheliae*
red-crowned *M. coronatus*
bob-tailed *Brachycope anomala*
Bishop, golden-backed *Euplectes aureus*
yellow-crowned *E. afer*
red *E. orix*
black-winged *E. hordeaceus*
orange *E. franciscanus*
Zanzibar *E. nigroventris*
black *E. gierowii*
fire-fronted *E. diadematus*
yellow *E. capensis*
Widowbird, long-tailed *E. progne*
Jackson's *E. jacksoni*
red-collared *E. ardens*
yellow-shouldered *E. macrourus*
buff-shouldered *E. psammocromius*
marsh *E. hartlaubi*
white-winged *E. albonotatus*
fan-tailed *E. axillaris*
Quelea, red-billed *Quelea quelea*
red-headed *Q. erythrops*
cardinal *Q. cardinalis*

FAMILY: ESTRILDIDAE (WAXBILLS, MANNIKINS,
TWINSPOTS, FIREFINCHES, ETC.)

Subfamily: Estrildinae

TRIBE: ESTRILDINI

Waxbill, black-faced *Estrilda nigriloris*
black-cheeked *E. erythromotos*
orange-cheeked *E. melpoda*
lavender *E. caerulescens*
cinderella *E. thomensis*
common *E. astrild*
black-rumped *E. troglodytes*
crimson-rumped *E. rhodopyga*
fawn-breasted *E. paludicola*
red-rumped *E. charmosyna*
black-tailed *E. perreini*

black-crowned *E. nonnula*
black-headed *E. atricapilla*
swee *E. melanotis*
anambra *E. poliopareia*
black-lored *E. nigriloris*
yellow-bellied *E. quartinia*
Abyssinian *E. ochrogaster*
Kandt's *E. kandti*
Firefinch, pale-billed *Lagonosticta landanae*
black-bellied *L. rara*
black-faced *L. vinacea*
black-throated *L. larvata*
red-billed *L. senegala*
bar-breasted *L. rufopicta*
African *L. rubricata*
Jameson's *L. rhodopareia*
brown *L. nitidula*
Mali *L. virata*
Reichenow's *L. umbrinodorsalis*
Negrofinch, grey-headed *Nigrita canicapilla*
pale-fronted *N. luteifrons*
chestnut-breasted *N. bicolor*
white-breasted *N. fusconota*
Pytilia, green-winged *Pytilia melba*
orange-winged *P. afra*
red-winged *P. phoenicoptera*
red-faced *P. hypogrammica*
lineated *P. lineata*
Twinspot, brown *Clytospiza monteiri*
dusky *Euschistospiza cinereovinacea*
Dybowski's *E. dybowskii*
Peters's *Hypargos niveoguttatus*
pink-throated *H. margaritatus*
green-backed *Mandingoa nitidula*
Seedcracker, black-bellied *Pyrenestes ostrinus*
lesser *P. minor*
crimson *P. sanguineus*
Bluebill, red-headed *Spermophaga ruficapilla*
Grant's *S. poliogenys*
western *S. haematina*
Crimson-wing, red-faced *Cryptospiza reichenovii*
Abyssinian *C. salvadorii*
Shelley's *C. shelleyi*
dusky *C. jacksoni*
Cordonbleu, blue-breasted *Uraeginthus angolensis*
red-cheeked *U. bengalus*
blue-capped *U. cyanocephalus*
Grenadier, purple *U. ianthinogaster*
common *U. granatina*
Waxbill, zebra *Amandava subflava*
Avadavat *A. amanadava*
Cut-throat *Amadina fasciata*
Finch, red-headed *A. erythrocephala*
Locustfinch *Ortygospiza locustella*
Quailfinch, African *O. atricollis*
red-billed *O. gabonensis*
Munia, bronze *Lonchura cucullata*
black & white *L. bicolor*
magpie *L. fringilloides*
brown-backed *L. nigriceps*
Silverbill, African *L. cantans*
grey-headed *L. griseicapilla*
Finch, cuckoo *Anomalospiza imberbis*
Waxbill, orange-breasted *Sporaeginthus*

subflavus
Antpecker, Jameson's *Parmoptila rubrifrons*
Woodhouse's *P. woodhousei*
Oliveback, grey-headed *Nesocharis capistrata*
white-collared *N. ansorgei*
Fernando Po *N. shelleyi*

Indigobird, village *Vidua chalybeata*
variable *V. funerea*
dusky *V. purpurascens*
Baka *V. larvaticola*
Jambandu *V. raricola*
twinspot *V. codringtoni*
Wilson's *V. wilsoni*
Whydah, straw-tailed *V. fischeri*
pin-tailed *V. macroura*
steel-blue *V. hypocherina*
shaft-tailed *V. regia*
Widowfinch, violet *V. incognita*
Paradise whydah, long-tailed *V. interjecta*
Togo *V. togoensis*
broad-tailed *V. obtusa*
eastern *V. paradisaea*
northern *V. orientalis*
Indigobird, pale-winged *Hypochera nigerrima*
Nigerian *H. nigeriae*

Chaffinch *Fringilla coelebs*
blue *F. teydea* **R**
Brambling *F. montifringilla*
Greenfinch, European *Carduelis chloris*
Goldfinch *C. carduelis*
Siskin, Eurasian *C. spinus*
Linnet, Eurasian *C. cannabina*
Warsangli *C. johannis* **R**
Redpoll, common *C. flammea*
Finch, crimson-winged *Rhodopechys sanguinea*
trumpeter *R. githagineus*
Bullfinch, Eurasian *Pyrrhula pyrrhula*
Finch, oriole *Linurgus olivaceus*
Rosefinch, common *Carpodacus erythrinus*
pale *C. synoicus*
Hawfinch *Coccothraustes coccothraustes*
Serin, red-fronted *Serinus pusillus*
European *S. serinus*
Syrian *S. syriacus*
citril *S. citrinella*
Ankober *S. ankoberensis*
Salvadori's *S. xantholaemus*

Canary, Cape *S. canicollis*
yellow-fronted *S. mozambicus*
forest *S. scotops*
brimstone *S. sulphuratus*
yellow *S. flaviventris*
white-throated *S. albogularis*
protea *S. leucopterus*
black-throated *S. atrogularis*
black-headed *S. alario*
Kenya grosbeak *S. buchanani*
Damara *S. leucolaema*
papyrus *S. koliensis*
white-bellied *S. dorsostriatus*

Abyssinian grosbeak *S. donaldsoni*
black-faced *S. capistratus*
Siskin, Drakensberg *S. symonsi*
Cape *S. tottus*
Abyssinian *S. nigriceps*
Seed-eater, black-eared *S. mennelli*
lemon-breasted *S. citrinipectus*
streaky-headed *S. gularis*
west African *S. canicapillus*
Abyssinian yellow-rumped *S. xanthopygius*
southern yellow-rumped *S. atrogularis*
Kenya yellow-rumped *S. reichenowi*
white-rumped *S. leucopygius*
yellow-throated *S. flavigula*
Reichard's *S. reichardi*
brown-rumped *S. tristriatus*
thick-billed *S. burtoni*
streaky *S. striolatus*
Principe *S. rufobrunneus*
Kipengere *S. melanochrous*
yellow-browed *S. whytii*
Citril, African *S. citrinelloides*
east African *S. hypostictus*
western *S. frontalis*

Subfamily: Emberizinae

Yellowhammer *Emberiza citrinella*
Bunting, cirl *E. cirlus*
rock *E. cia*
house *E. striolata*
cinnamon-breasted *E. tahapisi*
cinereous *E. cinceracea*
ortolan *E. hortulana*
Cretzschmar's *E. caesia*
rustic *E. rustica*
little *E. pusilla*
reed *E. schoeniclus*
black-headed *E. melanocephala*
Cabanis's *E. cabanisi*
African golden-breasted *E. flaviventris*
Cape *Emberiza capensis*
lark-like *E. impetuani*
Somali golden-breasted *E. poliopleura*
brown-rumped *E. affinis*
Socotra *E. socotrana*
Longspur, Lapland *Calcarius lapponicus*
snow *Plectrophenax nivalis*
corn *Miliaria calandra*
Crossbill *Loxia curvirostra*
Canary, São Tomé *Nesospiza concolor*
Grosbeak, golden-winged *Rhynchostruthus socotranus*

Index